Feminist Advocacy and Gender Equity in the Anglophone Caribbean

Routledge International Studies of Women and Place

SERIES EDITORS: JANET HENSHALL MOMSEN AND JANICE MONK, UNIVERSITY OF CALIFORNIA, DAVIS AND UNIVERSITY OF ARIZONA, USA

1. Gender, Migration and Domestic Service
Edited by Janet Henshall Momsen

2. Gender Politics in the Asia-Pacific Region
Edited by Brenda S.A. Yeoh, Peggy Teo and Shirlena Huang

3. Geographies of Women's Health
Place, Diversity and Difference
Edited by Isabel Dyck, Nancy Davis Lewis and Sara McLafferty

4. Gender, Migration and the Dual Career Household
Irene Hardill

5. Female Sex Trafficking in Asia
The Resilience of Patriarchy in a Changing World
Vidyamali Samarasinghe

6. Gender and Landscape
Renegotiating the Moral Landscape
Edited by Lorraine Dowler, Josephine Carubia and Bonj Szczygiel

7. Maternities
Gender, Bodies, and Spaces
Robyn Longhurst

8. Gender and Family among Transnational Professionals
Edited by Anne Coles and Anne-Meike Fechter

9. Gender and Agrarian Reforms
Susie Jacobs

10. Gender and Rurality
Lia Bryant and Barbara Pini

11. Feminist Advocacy and Gender Equity in the Anglophone Caribbean
Envisioning a Politics of Coalition
Michelle V. Rowley

Also available in this series:

Full Circles
Geographies of Women over the Life Course
Edited by Cindi Katz and Janice Monk

'Viva'
Women and Popular Protest in Latin America
Edited by Sarah A. Radcliffe and Sallie Westwood

Different Places, Different Voices
Gender and Development in Africa, Asia and Latin America
Edited by Janet Momsen and Vivian Kinnaird

Servicing the Middle Classes
Class, Gender and Waged Domestic Labour in Contemporary Britain
Nicky Gregson and Michelle Lowe

Women's Voices from the Rainforest
Janet Gabriel Townsend

Gender, Work and Space
Susan Hanson and Geraldine Pratt

Women and the Israeli Occupation
Edited by Tamar Mayer

Feminism / Postmodernism / Development
Edited by Marianne H. Marchand and Jane L. Parpart

Women of the European Union
The Politics of Work and Daily Life
Edited by Maria Dolors Garcia Ramon and Janice Monk

Who Will Mind the Baby?
Geographies of Childcare and Working Mothers
Edited by Kim England

Feminist Political Ecology
Global Issues and Local Experience
Edited by Dianne Rocheleau, Barbara Thomas-Slayter, and Esther Wangari

Women Divided
Gender, Religion and Politics in Northern Ireland
Rosemary Sales

Women's Lifeworlds
Women's Narratives on Shaping Their Realities
Edited by Edith Sizoo

Gender, Planning and Human Rights
Edited by Tovi Fenster

Gender, Ethnicity and Place
Women and Identity in Guyana
Linda Peake and D. Alissa Trotz

Feminist Advocacy and Gender Equity in the Anglophone Caribbean

Envisioning a Politics of Coalition

Michelle V. Rowley

Routledge
Taylor & Francis Group
New York London

First published 2011
by Routledge
711 Third Avenue, New York, NY 10017

Simultaneously published in the UK
by Routledge
2 Park Square, Milton Park, Abingdon, Oxon OX14 4RN

First issued in paperback in 2013
Routledge is an imprint of the Taylor & Francis Group, an informa business

© 2011 Taylor & Francis

The right of Michelle V. Rowley to be identified as author of this work has been asserted by her in accordance with sections 77 and 78 of the Copyright, Designs and Patents Act 1988.

Typeset in Sabon by IBT Global.

All rights reserved. No part of this book may be reprinted or reproduced or utilised in any form or by any electronic, mechanical, or other means, now known or hereafter invented, including photocopying and recording, or in any information storage or retrieval system, without permission in writing from the publishers.

Trademark Notice: Product or corporate names may be trademarks or registered trademarks, and are used only for identification and explanation without intent to infringe.

Library of Congress Cataloging-in-Publication Data
Rowley, Michelle V.
 Feminist advocacy and gender equity in the Anglophone Caribbean : envisioning a politics of coalition / by Michelle V. Rowley.
  p. cm. — (Routledge international studies of women and place ; 11)
 Includes bibliographical references and index.
  1. Feminism—Caribbean, English-speaking. 2. Women's rights—Caribbean, English-speaking. 3. Women—Political activity—Caribbean, English-speaking. I. Title.
 HQ1525.4.R69 2011
 305.4209729'0917521—dc22
 2010029516

ISBN13: 978-0-415-84765-0 (pbk)
ISBN13: 978-0-415-87854-8 (hbk)
ISBN13: 978-0-203-83288-2 (ebk)

For
Merle DePradine,
Edgar "Tula" Rowley
(1919–1994),
and
Hilton Rowley

Contents

List of Illustrations		xi
List of Tables		xiii
List of Abbreviations		xv
Preface		xxi
Acknowledgments		xxvii
1	Mapping the Terrains of Gender Equity: Gender Mainstreaming, Contexts, Compromises, and Conflicts	1
2	Crafting Maternal Citizens: Historicizing Institutional Subjectivities within Gender Mainstreaming	20
3	Co-opting Gender and Bureaucratizing Feminism: Exploring Equity through the Institutionalization of "Gender"	55
4	Reproducing Citizenship: A 20/20 Vision of Women's Reproductive Rights and Equity	91
5	Keeping the Mainstream in Its Place: Sexual Harassment and Gender Equity in the Workplace	127
6	Development and Identity Politics: Securing Sexual Citizenship	171
Notes		205
Bibliography		221
Index		237

Illustrations

FIGURES

4.1	Timeline: Passage of Barbados Medical Termination of Pregnancy Act, 1983.	117
4.2	Timeline: Passage of Guyana Medical Termination of Pregnancy Act, 1995 (Act No. 7 of 1995).	118

MAPS

Map of the Caribbean   xix

Tables

3.1	Staffing Complement and Gender Training of Regional Units and Bureaus	64
3.2	Comparative HDI, GDI, and GEM Ranking for Selected Countries	70
4.1	Overview of Women's Reproductive Health	113
4.2	Abortion Rates at Queen Elizabeth Hospital, Barbados (1991–2001)	121
4.3	Abortion Rates, Guyana (1996–1999)	121
4.4	Maternal Mortality Rates, Trinidad and Tobago (1991–2001)	123
5.1	Comparison of Male and Female Strength and Attrition (2000–2004)	132
5.2	Application Levels and Examination Performance by Sex (2000–2004)	140
5.3	Male Wastage on Medical Grounds	145
5.4	Female Officers' Promotion in the RBPF (2000–2004)	149
5.5	Overview of Sexual Harassment Legislation in the Anglophone Caribbean	155

Abbreviations

ASPIRE	Advocates for Safe Parenthood: Improving Reproductive Equity
CAFRA	Caribbean Association for Feminist Research and Action
CAISO	Trinidad and Tobago Coalition Advocating for Inclusion of Sexual Orientation
CARICOM	Caribbean Community
CARIFLAGS	Caribbean Forum for Liberation and Acceptance of Genders and Sexualities
CASH	Coalition against All Forms of Sexual Harassment
CEDAW	Convention on the Elimination of All Forms of Discrimination Against Women
CIDA	Canadian International Development Agency
DAWN	Development Alternatives with Women for a New Era
ECOSOC	Economic and Social Council
FHH	Female-Headed Households
FPA	Family Planning Association
GAD	Gender and Development
GDI	Gender-related Development Index
GDP	Gross Domestic Product
GEM	Gender Empowerment Measure
GM	Gender Mainstreaming
GUYBOW	Guyana Rainbow Foundation
HDI	Human Development Index
HDR	Human Development Report

HRC	Human Rights Committee
IDB	Inter-American Development Bank
ICPD	International Conference on Population and Development
IGO	Intergovernmental Organizations
ILGA	International Lesbian and Gay Association
IMF	International Monetary Fund
ISER	Institute for Social and Economic Research (now SALISES)
J-FLAG	Jamaica Forum for Lesbians, All Sexualities and Gays
LGBT	Lesbian, Gay, Bisexual, and Transgender
MDC	Movement for Democratic Change
MDG	Millennium Development Goal
MSM	Men Who Have Sex with Men
NGO	Non-governmental organization
NWCSA	Negro Welfare Cultural and Social Association
OAS	Organization of American States
OECS	Organization of Eastern Caribbean States
PAHO	Pan-American Health Organization
RBPF	Royal Barbados Police Force
SASOD	Society against Sexual Orientation Discrimination
UGLABB	United Gay and Lesbian Association of Barbados
UN	United Nations
UNDP	United Nations Development Program
UNECLAC	United Nations Economic Commission for Latin America and the Caribbean
UNHCR	United Nations High Commissioner for Refugees
UNICEF	United Nations Children's Fund
UNIFEM	United Nations Development Fund for Women
U-RAP	University of the West Indies Rights Advocacy Project
UWI	University of the West Indies

UWIDITE	University of the West Indies Distance Teaching Experiment
WAD	Women and Development
WB	World Bank
WHO	World Health Organization
WICP	Women in the Caribbean Project
WID	Women in Development
WIRC	West India Royal Commission
ZANU-PF	Zimbabwe African National Union-Patriotic Front

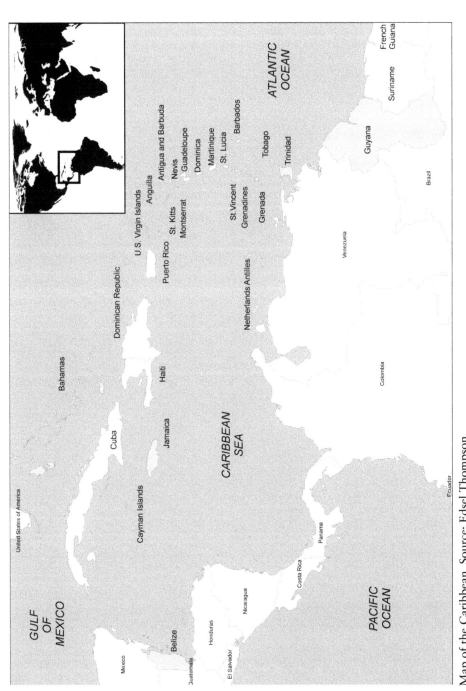

Map of the Caribbean. Source: Edsel Thompson.

Preface

> **People Power (The Leader Speaks)**
> In the name of the people
> I give you the People's Constitution
>   in which the rights of the people
> Have been enshrined.
> So now that the people's rights
> Are enshrined, meaning dead,
> Let us get on with the business
> Of building the nation
> For the good of the people.
> But remember: be vigilant.
> Anybody who trouble me trouble you
> For you is *me*; I am the people—
> In the name of the people
> And the people
> And the people.
>
> Edward Baugh, A Tale from the Rainforest

Jamaican poet Edward Baugh's "People Power" presents us with the vexing contradictions of speaking in the name of a collective "we" where citizenship becomes equated with a range of im/possibilities. Baugh captures these im/possibilities by first framing the liberating aspects of constitutional rights, but then he inverts "we" the collective and "I" the leader to show these liberating aspects to be hardly more than a hollow guarantee that really belong solely to "I" the leader. The emptiness of any constitutional guarantee reveals itself as the poem progresses, and the leader reveals himself to usurp and become the people (I am the people), thereby opening the space for the idea of the "People's Constitution" to collapse into the Leader's constitution. Here, Baugh invites us to reflect on the shallowness of these "rights" through the possible nuances of "enshrinement." Although "enshrinement" offers an umbrella of protection, on the one hand, it suggests, on the other hand, a kind of settling—stasis. If the latter is indeed the case, then the people's constitution by virtue of being the leader's constitution renders "people's rights" as irrevocably dead. Further, "People Power" invites us to consider the nature of what is being enshrined. If enshrinement signals death, and we mark such death as the erasure of the people and other forms of inequity, then how do we contend with political scenarios that make such enshrinement sacrosanct?

In this piece, Baugh confronts one of the ironies of nationalism: the invocation of "rights," rather than producing a liberating effect, can become entombed as little more than a rhetorical gesture and rendered as an impediment to "the nation's business." Here, it is possible that "the nation's business" symbolizes both the imperatives of the market and an exercise in daily affairs void of any attention to the people's "rights," in other words, "business as usual." Additionally, Baugh invites us to consider a leader whose identification with "the people" is deeply disturbing. His identification is disturbing in that his alignment with the "people" refuses to acknowledge the need for reciprocity. If the leader is troubled, then so are the people. But what of the people who are troubled? The leader's formulation endorses a unidirectional sense of responsibility in which homage to the leader insists upon the people's assuming personalized umbrage to the leader's offense. There is no such umbrage should the people be troubled; therefore, at its denouement, "and the people" fades into a repetitive, weakening refrain.

Feminist Advocacy and Gender Equity in the Anglophone Caribbean aims to fracture the "nation's business" and "the people." We know that "the people" in their varied forms of differentiation rarely have equal levels of access to either justice or community. Centering and expanding the call for gender equity and gender justice, *Feminist Advocacy* challenges, as does Baugh's "People's Constitution," the prevailing will of the leader over "the people," the ceding of social justice to economic business as the nation's primary business, and the ever-increasing marginalizing of disadvantaged constituencies from a conceptualization of "the good of the people." In this project, I push Baugh's use of the "People's Constitution" to examine further the implications of the use of "constitution" as body. While Baugh invokes "body" here as both a political and social collective, as well as a legal instrument that protects rights, *Feminist Advocacy* considers "the body" specifically as a form of physical embodiment—an embodiment that is in dialogue with Baugh's social and political collectives. Therefore, if we insert the body (of women, sexual minorities) into Baugh's progressing critique of this social and political collective, would this prompt us to consider whether the constitutional protection of embodied autonomy might also be hollowly enshrined?

Like many territories influenced by the goals and strategies of the Beijing Platform for Action, the state-based strategy of gender mainstreaming has been one of the central vehicles through which claims of equity have been made in the name of "the people" in the region. In *Feminist Advocacy*, my analysis builds on three interconnected terrains of discussion. First, my work examines the extent to which gender mainstreaming has been effective toward the realization of gender equity in the Anglophone Caribbean. Second, I explore the ways in which this strategy of gender equity has been unable to incorporate an embodied understanding of equity, thereby leaving un-addressed questions of women's reproductive health and bodily integrity. Finally, I call for a decentering of gender so that gender equity development strategies are better positioned to address the marginality of queer subjects within development debates.

I begin *Feminist Advocacy* by mapping the conceptual terrain of gender equity through a critique of gender-mainstreaming approaches. My country selection is guided by the extent to which various arms of civil society have engaged the issue under discussion. Consequently, *Feminist Advocacy* examines gender mainstreaming (Trinidad), reproductive rights (Barbados, Guyana, Trinidad and St. Lucia), sexual harassment (Barbados) and rights for sexual minorities (Guyana). I examine the confinements and possibilities of this technical approach by examining the ways that gender mainstreaming encounters new conceptual and operational hurdles as it moves from one tier of implementation to another. I address my concerns about gender mainstreaming's effectiveness in achieving gender equity in Chapters 2 and 3. Chapter 2 begins with a post-development critique of developmentalism that drives hegemonic assumptions about the "South" as a geo-political fiction.[1] Using this post-structural critique of development, I argue that prescriptive, ethnocentric assumptions of "gender" have accompanied the United Nations' (UN) global dissemination of mainstreaming as a strategy of gender equity and that these prescriptions have the capacity to limit rather than enable the local relevance of gender mainstreaming as a strategy of equity. The neoliberal impulses that inform contemporary development debates have successfully written out the importance of any kind of "historical analyses"; here, gender mainstreaming as a strategy of development is no exception. Consequently, in the same way that a country's "poverty" or Human Development Index (HDI) ranking now emerges as a *condition or identity* without antecedents of under-development and easily assumed to be rectified by incorporation into the market, neoliberal renderings of "gender" similarly present an understanding of gender in which gendered subjects are first and foremost economic actors, who once appropriately trained (whether through atypical professions or micro-enterprise) and incorporated into the labor market, have the capacity to disrupt a wide range of existing inequities.[2] Such an approach, I suggest in this chapter, does not acknowledge the multiple ways in which individuals experience their gender identity and, more importantly, does not account for how one's gendered subjectivity, even as a "worker," has been historically and discursively produced within the local.

Chapter 2, therefore, offers the reader a methodological excursion into what I refer to as *gender-as-genealogy* as a way of showing why local histories of gender relations matter to contemporary mainstreaming approaches to gender equity. In this chapter, I examine how the Trinidadian state has historically privileged a sacrificial model of women's maternal identities in the provision of welfare services. Through this historicization, I point to the ways in which state policy and state agents function to encourage a performance of the mother-as-Madonna trope, a performance, I maintain, that is informed by colonial and nationalist antecedents. This examination lays the ground for a more in-depth understanding of gender as a historically structured set of relations that are always-already at work *alongside of* the prescriptive understandings of gender that accompany gender mainstreaming approaches. I tell a particular story around these set of relations for I think it important to address (if only conceptually) these antecedents because, as I suggest throughout this project, the resulting tropes recur in a wide variety of policy debates that are pivotal to contemporary mainstreaming ideals (e.g., the maternal trope is deployed in arguments against reproductive rights). Further, they foreclose on other subject positionalities for women and stand as the subtext for how state-managers understand the "appropriate" frame of reference for "gender policy."

In Chapter 3, I continue my examination of gender mainstreaming's effectiveness as a strategy of equity by exploring the systemic and discursive hindrances to gender mainstreaming within the national machinery in Trinidad.[3] This discussion exemplifies the obstacles faced by feminists in bureaucratic structures—or "femocrats," to draw on sociologist Hester Eisenstein's phrasing—when given responsibility for mainstreaming gender in hostile conditions.[4] For example, I argue that bureaucratic cultures by virtue of their dispassionate, detached nature stand as antithetical to the politics of gender mainstreaming. In analyzing the experiences of female ministers of government and technocrats with responsibility for gender mainstreaming, I examine the contradictory ways in which narratives of "illegality," "madness," and "hysteria" are used within the supposed dispassionate bureaucratic structure to undermine the credibility of the technocratic change agents and, by extension, their agenda. According to both of these renderings, the bureaucratic ethos and practices of the parliamentary system of government in the Caribbean have not provided a supportive environment for gender mainstreaming or for women's advancement. In other words, the ethos and practices are such that they have formulated many *legitimate* ways of saying "no" to women's equality. The rhetorical strategies at work within national machineries produce a sleight of hand in which "gender" is discussed as relevant to questions of equity but ignored in processes of implementation; apparently central, while woefully marginal. The data for this chapter was collected in 2002; so, although things may have changed in the day-to-day running of the Gender Affairs Bureau of Trinidad and Tobago, I offer it as an exemplary analysis of the dilemmas faced by many "femocrats" throughout the region.

In Chapter 4, "Reproducing Citizenship: A 20/20 Vision of Women's Reproductive Rights & Equity," I begin my engagement with *Feminist Advocacy*'s second objective, namely, an exploration of how gender mainstreaming fails as a strategy of gender equity when called upon to account for women's bodies and their bodily integrity. I examine the ongoing attempts by one feminist organization, Advocates for Safe Parenthood: Improving Reproductive Equity (ASPIRE), to secure the passage of termination legislation in Trinidad and Tobago and St. Lucia. I explore the methodologies and political interface of ASPIRE with state-managers on the abortion debate. Struck by the reproductive conservatism that informs mainstreaming approaches to equity, I highlight how this conservatism finds mirroring reinforcements at both national and international spheres of operation. I also delineate the processes that informed the passage of termination legislation in Barbados and Guyana as a means of showing the ways in which women's national machinery benefits from processes of coalition that occur outside of its framework while at times remaining peripheral to the action and processes of change.

"Keeping the Mainstream in Its Place: Sexual Harassment and Gender Equity in the Workplace," Chapter 5, continues my interrogation of the ways bodily integrity continues to be a sticking point in the achievement of gender equity in the region. I pose the following questions: What kinds of feminist models of organizing can we envision for the twenty-first century in a region characterized by the rhetoric of reverse gender discrimination? And, to what extent have commitments to gender equity made it possible for women to lay claims to greater levels of sexual autonomy within the public sphere? To respond to these questions, I examine the advocacy work done by the Coalition against All Forms of Sexual Harassment (CASH) toward the passage of sexual harassment legislation in Barbados. This chapter builds its analysis on a case study on gender equity in the work place done with the Royal Barbados Police Force (RBPF). What becomes clear in this chapter is that it is not enough to merely incorporate women into the workplace as the primary goal of gender equity. I show that women once incorporated as "workers" within the labor market must, in turn, find a range of rhetorical and operational strategies to maintain their bodily integrity and circumnavigate marginalizing practices—inclusive of harassment.

Chapter 6, "Development and Identity Politics: Securing Sexual Citizenship," intervenes in the region's self-aggrandized constellation of "Visions"—or strategic development plans—that now punctuate the political rhetoric. In this chapter, I engage in a disciplinary dialogue with the field of development studies in order to gain a greater degree of queer visibility within development debates. This chapter encourages the reader to engage more substantially in an "ethics of development" that critiques the extent to which sexual and gender minorities are silenced by this rhetoric of developmentalism. In this chapter, I focus my attention on a reading of the Yogyakarta Principles (2007), which aim to apply international rights law to questions of sexual orientation and gender identity and bring these Principles into a speculative, exploratory dialogue with extant literature on Caribbean sexuality. Finally, I explore the ways in which sexual dissidents emerge as a "fetish" within the political landscape and focus on a practice of what I refer to as symbolic capture as a mode of activism that challenges the ways in which queer subjects have been over-determined or foreclosed within the development and political landscape of the region. To this end, I provide a preliminary overview of the work being done by Society against Sexual Orientation Discrimination (SASOD) in Guyana as a foray into the importance of queer symbolic capture in transforming the equity terrain of the Anglophone Caribbean.

I hope that in the chapters that follow I prompt my (activist) reader to rethink the extent to which national machineries might be the appropriate locus of achieving gender justice and equity in the Anglophone Caribbean. It is my wish to engage the conceptual limitations and possibilities inherent in the discursive circulation of gender mainstreaming. Here, somewhat provocatively, I invite the reader to consider whether the category "gender" stands as the means by which equity and justice can be achieved in the region, and if so, how? Throughout the project, I chart where possible the ways that the Caribbean women's movement and activism generally is refashioning itself, at times consciously, other times through expediency, into an organization that is more diffuse, less centralized, and less dependent on unitary points of leadership. In other words, I optimistically see a postmodern organizational formation at work that aims to address the global political climate where social issues and discriminatory practices exist on multiple terrains simultaneously. I very deliberately re-position the body as critical to questions of equity by engaging issues of sexual harassment, reproductive rights, and rights for sexual minorities. I am not suggesting, as one reader has challenged, that to be feminist one should be on a particular side of the reproductive debates or sexual minority rights. My own position will be clear: that sexual minority rights and women's reproductive health are pivotal markers to framing a just society. That said, although I do not ascribe any one political position to feminist debates or a political position to anyone, I am sure that to be apolitical and not engage these debates would be a myopic option for Caribbean feminists.

Acknowledgments

This project has been many years in the making and has benefitted from the support and good cheer of individuals situated across a number of different institutions and geographical locations. The background story to *Feminist Advocacy* is truly transnational. I first learned that I could think politically about women because of the efforts of the Women and Development Studies Group and the Centre for Gender and Development Studies, University of the West Indies. I continue to consider my time at Clark University a pivotal period in my intellectual development because of colleagues and friends such as Elora Chowdhury, Gai Liewkat, Young Rae Oum, and Barbara Schulman. Long Live PITA! I continue to remain grateful for the hard work and sustained belief that Cynthia Enloe, Barbara Thomas-Slayter, and Kiran Asher have placed in me.

I am indebted to my former colleagues at the University of Cincinnati who remain some of the most welcoming people I know. I want to especially thank Deb Meem, Michelle Gibson, and Anne Sisson-Runyan for their ongoing friendships and intellectual support on this journey.

I am presently in a community of scholars at the University of Maryland whose institutional activism and commitments to social justice provide a sense of daily optimism. I often say that, beyond my doctoral committee at Clark, these folks have had the most to do with my early doctoral development, and I now feel truly privileged to have the collegial support of Lynn Bolles, Elsa Barkley-Brown, Bonnie Thornton-Dill, Seung-kyung Kim, Katie King Jeffrey McCune, Tara Rodgers, and Deborah Rosenfelt. Claire Moses and Ruth Zambrana guided my early years here at Maryland in ways that continue to warm my heart. To my larger circle of community scholars, Nancy Struna, Psyche Williams-Forson, Hilary Jones, Faedra Carpenter, and Christina Hanhardt, I thank you for every meal, every cup of tea, and every end of semester commiseration. Hands down and without reservation, nothing that I do on a daily basis would be possible without Laura Nichols, Annie Carter, and Cliffornia Howard. I cannot thank you enough for your support.

At times, completing this project often seemed impossible. I would like to extend my appreciation to those who not only insisted that the project was possible but also constantly believed in my ability to bring it to fruition. None of this would be possible without friends who are willing to be my friend in absentia, as I occasionally disappear for purposes of intellectual self-immolation. Thanks to Wayne Allard, Lynette Joseph-Brown, Donette Francis, Rhonda Denise Frederick, Roanna Gopaul, Roy McCree, Tracy Robinson, Patricia Joan Saunders, Meera Sehgal, Marjorie Thorpe, Amar Wahab, and Leighton Waterman.

Any weaknesses of reasoning and thinking are my own, but I must acknowledge that this work has been made better by the intellectual feedback offered over time by Cynthia Enloe, Claire Moses, Jane Parpart, Tracy Robinson, Faith Smith, Amar Wahab, and various anonymous readers. I also want to thank Ms. Roberta Clarke, Regional Program Director, for her permission to use data from my previously commissioned research on gender mainstreaming in the region (United Nations Development Fund for Women [UNIFEM]/Rowley 2003), the CASH for permission to use data gathered from my UNIFEM-funded research on sexual harassment and gender equality in the RBPF, and Prof. Eddie Baugh and Sandberry Press for their permission to use the wonderful poem "People Power (The Leader Speaks)" that begins the book.

Thank you to my family, particularly my mother who, like my father, to this day continues to stand behind me with a very simple admonition, "Do your best, try your best." You all remind me that life will always extend beyond academia. I love the fact that you don't really care about what I do. To my always little ones, some of whom are now taller than I, this project has been written in anticipation of a future Caribbean that you do not yet understand as important but that I hope you will one day work to achieve. Ellen, I could not have finished this without your support, unwavering enthusiasm, and critical eye. This book is as much yours as it is mine. Thank you for lifting me over my sentences. Shelby, this is for you. Had you not stood under my balcony to sing me out of my funk some twenty years ago, who knows where I'd be. I know you're smiling . . . wherever you are.

1 Mapping the Terrains of Gender Equity
Gender Mainstreaming, Contexts, Compromises, and Conflicts

> To be useful requires communities ... I cannot write without an (at least imaginary) audience—a community. Communities provide one with puzzles to address; theoretical frames to appropriate; purposes to evaluate, adopt, or reject; conversation partners to engage; multiple senses of identity; loyalties that can moor and enrich one's sense of place; and a sharp brake against one's narcissism and grandiosity. In defining boundaries and identities, communities name and constitute outsiders and enemies who it is necessary and correct to attack. They also provide one with reasons to continue to slog through the mud, especially when the intrinsic satisfaction of the task appears slight. One can at least have the fantasy that one is acting on behalf of something larger and more important than oneself.
>
> Jane Flax, *Disputed Subjects: Essays on Psychoanalysis, Politics and Philosophy*

This project emerges out of a passion for community and the need to map the ways gender equity has been achieved by or at times eluded feminists in the Anglophone Caribbean.[1]

I write with urgency—an urgency that prompts me to intervene philosophically and theoretically in what seems to be an increasing sense of apathy regarding matters of equity. To understand this apathy requires that we map the persistent political unwillingness of state-managers to aggressively pursue an agenda of equity for marginalized citizens in the region. From this point, we would need to traverse along the (shrinking) cache of civil resources available for challenging practices of discrimination within civil society. Our mapping would then turn our attention to the politics of a contemporary global economy that is increasingly hostile to small island-state economies, thereby increasing the difficulty with which feminist activists are able to stem the trade off of social concerns for economic exigencies. And, in the hills and gullies of this cartography, we would need to explore the politics of feminist advocacy in the Anglophone Caribbean. As we pause here, we might be compelled to ponder the extent to which previous strategies of gender equity served to produce, even if unwittingly, their own marginal subjectivities. Have feminists undermined their own political agenda by offering state-managers very limited resources with which to think through the complexity of the two categories that inform the concerns of this book—"gender" and "equity?"[2] Further, what are the new terrains that we must map if claims toward gender equity are to matter as part of the region's understanding of itself in the twenty-first century? In *Feminist Advocacy and Gender Equity in the Anglophone Caribbean*, I engage these questions, using a number of different strategies and modes of engagement.

UNDER WESTERN GUISE—POST-DEVELOPMENT'S UNVEILING

Fully cognizant of feminist concerns about post-structuralism, I nonetheless ground *Feminist Advocacy* in what I see as post-structuralism's most transformative contributions within development debates: its sustained engagement with historicizing contemporary inequity, as well as its interrogation of development discourses. However, throughout *Feminist Advocacy*, I inform this post-structuralist critique with a feminist sensibility that sends us in search of the material effects of discourse.

To say that post-structuralism has influenced the shape of development debates does not quite capture the shifts that have occurred over the past two decades. Post-structuralism's influence has been so profound that the very idea of "development" has been irrevocably called into question within a "post-development" framework. An overview of the increasing number of post-development readers on the market makes it clear that post-development's defining feature is not an inextricable alliance with post-structuralism (since its advocates are by no means all post-structuralists); rather, its scathing critique of the stranglehold that development has had in structuring global symbolic hierarchies and the psyche of individuals who live with and through these inequities.

Anthropologist Arturo Escobar stands at the forefront of this move with his insightful discussion of development's regulating force. His work denaturalizes the idea of development as desirable and Messianic and explores the rhetorical, symbolic, and material processes that make "development" a prominent feature within our social imaginary (1984, 1995). Drawing on a discourse analysis, Escobar carefully analyzes the consolidation of the idea of "development" and the regimes of representation that result through the twin processes of institutionalization and professionalization (1995, 17). The "professionalization" of development marks the process through which we encounter the disciplinary naming of the Third World as a problem (e.g., economic poverty, social dysfunction, and environmental liability). It follows that one can only address these disciplinary realizations through institutional and systemic policy and programming responses, hence, the processes of "institutionalization."

Escobar challenges these movements by invoking development as a discourse. In so doing, he places analytical emphasis on "the process through which social reality comes into being . . ."; that is to say, the creation of, through multiple fields of operation, a "space in which only certain things could be said and even imagined . . ." (ibid., 36). His critique strips the twin processes of institutionalization and professionalization of their benign aura and reveals the hierarchies and symbolic and material economies that they instantiate. As such, his work interrogates the dominant epistemological framework of development and opens a space for different epistemological articulations.

Feminist criticism of post-structural interventions in development debates are not without merit. Among feminist concerns was the need to distinguish post-structuralism's rhetoric about difference from a feminist politics of difference as a practice of material inclusion (Bordo 1990). For some, the predominance of white, male theoreticians, and the "inaccessibility" of the theoretical language undermined an egalitarian agenda. Post-structuralism, it was felt, hinged on narrowly defined "membership," and fed an acritical embrace of another masculinist discourse by feminists (ibid.). Post-structuralism's rendering of subject formation as contradictory and fragmented further presented a crisis of sorts for the identity categories critical to feminist political action. Many liberal and Marxist feminists questioned whether such an analysis did not simultaneously undermine the relevance of categories such as race, gender, class, sex, and discriminatory structures of racism and patriarchy (Walby 1990, cited in Parpart). How would feminists be able to talk about women's resistance, agency, and experience—categories foundational to feminist criticisms—if there were no subject?

For Susan Bordo, sidestepping the relevance of these analytical categories would only serve to maintain the status quo. She writes:

> It is no accident, I believe, that feminists are questioning the integrity of the notion of a "female reality" just as we begin to get a foothold in those professions which could be most radically transformed by our (historically developed) Otherness and which have been historically most shielded from it. (151)

Here, Bordo captures the ultimate expression of feminist disaffection: the belief that skepticism for the primacy of gender would somehow encourage feminists to act against their own intellectual and political interests. Admittedly, Bordo commends deconstruction for its ability to unbind the ties that limit human possibility. Yet she, as did many others, remains less than convinced of its transformative potential (ibid., 153).

Despite Bordo's disavowal, I think an engagement with post-structuralism has much to offer the pursuit of gender equity. A consideration of development as discourse does not preclude the possibility for resistance, agency, or a productive engagement with the category of experience. Post-development scholars' exploration of development's regulatory regimes does not preclude a discussion of how those marginalized constituencies may participate in counter-discourses to these regimes of representation. As such, postdevelopment scholars, as I do here, pursue examinations of counter-discourses despite our awareness that the resistive practices are "tenuous at best" (Escobar 1995, 223). Wendy Harcourt and Arturo Escobar in their introduction to *Women and the Politics of Place* (2005) call for a critical study of the politics of place. They consciously focus their attention on the transformative politics of place, a politics which they note ". . . may involve resistance, but it also involves re-appropriation, reconstruction, reinvention, even relocalization of places and place based practices" (ibid., 3).

I am persuaded here by what will be a recurring perspective in this work, namely, that we cannot know what resistance will look like a priori without consideration of place, context, and the specific exercise of power. In this context, "resistance" and "agency" are not already named as class, race, or gender but remain contextual and contingent. This leaves it open for both reader and actor to consider the ways in which transformation may possibly be arrived at even through forms of complicity. Such an approach, for example, is reflected in Escobar's observation that there can be no pure space of resistance; rather, what we are constantly reading for are the ways in which resistance models are formed as "complex hybridizations within dominant models" (1995, 96).

Consequently, agency should be seen as a form of re-signification. This is a very different process from a prescriptive sense, for example, of what we expect women to do in situations of domestic violence, in response to the state, or in compliance with mainstreaming policies. The end result of re-signification may, at times, indeed, look like what we "expect" women, sexual minorities, or people of color "to do" in the face of oppression. However, re-signification opens a wider range and deeper nuance of what agency might look like. I am building here on Butler's argument that rather than a free wielding "I," what we do have is an "I" that is always in dialogue with dominant norms and practices.

Agency as resignification captures Butler's understanding that we are in some way "able to do something with what is done to me." Agency as re-signification is, therefore, an examination of how individuals respond to the ways they may be rendered or held hostage by dominant representations or politics in any given society. Analyzing agency through processes of re-signification also acknowledges that there may be systemic constraints that are so overwhelming that they mask or make certain identities completely unintelligible (2004, 3).

In *Feminist Advocacy*, a post-structural critique informs my interest in understanding *how* and *why* some categories have come to matter in our understanding of gender equity, to the exclusion of others. It informs my methodological approach, which calls for a genealogical understanding of gender within local spaces of practice. It guides my discussion of which bodies become hypervisible or made to be spectacles. And, it prompts an examination of how the category "gender" is deployed within contemporary equity debates—in which gender, like balance of payments, becomes another thing that the countries in the global South need to "get right."

GENDER MAINSTREAMING—A QUEST FOR GENDER EQUITY?

In *Feminist Advocacy*, I see gender mainstreaming as a project, despite its variable success, that is inherently committed to the realization of equity. I use the term "equity" rather than "equality" as a way of articulating my own utopian vision for the region. I emphasize conditions of *equity* because conceptually it is more responsive to the differential experiences that subjects bring to the table and opens a discussion for subjects' intersectional identities in ways that the rather static notion of equality (which often deploys singular and "fixed" subject markers—e.g., *race* equality, *gender* equality) does not. My use of "equity" presents some difficulty in my discussion because "equality" discourses are the preferred ones within the legal framework that I reference in my analysis. However, where appropriate I will distinguish which concept is being used.

My choice is informed by social economist Naila Kabeer's observation that an equity approach is one which "requires the transformation of the basic rules, hierarchies, and practices of public institutions" (1994, 87). Kabeer's approach to equity clarifies the advantages of "equity" over "equality," distinguishing between one's access (opportunity) and whether such access enhances one's ability to live with similar possibilities as others do (outcome). She illustrates this distinction anecdotally with the fable of the fox and crane, which are both invited to dinner and fed equally out of a saucer. Despite the similarity of access, the crane's long beak makes her ill-equipped to optimally enjoy the meal presented before her. I use "equity" here to insist on the systemic changes that may be necessary for a person to live optimally without having to simultaneously denigrate or mar her inherent differences.

Concerns that the language of equality also conveys a Western, individualistic logic have prompted some feminists to question the applicability of these approaches in culturally different locations. In her article "Women's Rights, CEDAW and International Human Rights Debates toward Empowerment?" (Convention on the Elimination of All Forms of Discrimination against Women), feminist legal scholar Shaheen Ali questions the efficacy that claims to formal equality hold for women generally. She notes that "equality" fails to deal with women's specific differences, that its Enlightenment origins offer women little more than a masculine standard of achievement, and as I have argued here, that once individuals begin from unequal positions, they are unlikely to achieve equality as a result of legislation that has this as its goal (2002, 73). My use of equity works similarly to Ali's call for "substantive equality," which she states would "require those in authority (i.e., men and the state) to accept responsibility for fulfilling the material needs of women and children and other disadvantaged sections of society in their charge and for providing them with access and control over resources (ibid., 75).

Both as strategy and concept, gender mainstreaming stands as the primary touch point in discussions of gender equity within the region's national machinery. It calls for the deployment of state resources, the allocation of personnel, and the reworking of operational procedures. If taken seriously, its operationalization demands a radical break with the ways in which the region addresses issues of women's advancement, justice, and equity. However, its very location within state machinery often leads to the creation of its own incapacities.

GENDER MAINSTREAMING—FEMINIST FRIEND OR FOE?

Gender mainstreaming is widely touted as the technical instrument necessary for achieving gender equity within nation-states. The Beijing Declaration and Platform for Action (Beijing Platform), emanating from the UN Fourth World Conference on Women in 1995 and adopted by all participating members, highlighted the importance of strengthening national machinery and developing the framework, methods, and tools for incorporating gender into policy making and planning, including macro-level long-term plans (e.g., five-year strategic plans). Integral to most definitions of gender mainstreaming are the achievement of gender equity through institutional and legislative changes, an ongoing critique and analyses of socioeconomic and political asymmetries, and the development of monitoring and evaluative mechanisms to facilitate sustainable implementation.

It is now accepted practice to cite the Beijing Platform for Action as the driving engine for the global dissemination of gender mainstreaming approaches (Beal 1998; True 2003; Carney 2003). While the Beijing Platform remains unquestionably pivotal in establishing and disseminating a working template of conditions, strategies, and goals for gender mainstreaming,[3] resolutely adhering to this originary narrative hides three important features that lend to gender mainstreaming's tenacity. First, gender mainstreaming, as discussed within the Beijing Platform, represents a culmination of many decades of strategic action and analytic shifts regarding women's relationship to national and global resources. Second, the Beijing Platform's template for gender mainstreaming remains responsive to narrow, more "palatable" definitions of gender—hence, a kind of innocuous sustainability. Third, there are now multiple points of institutional influence that set about to name and describe gender mainstreaming processes.

The pre-Beijing life of gender mainstreaming is evident from as early as the first UN Decade for Women (1976–1985), which led to governments' establishing women's desks, departments, and bureaus toward the "advancement" of women's concerns into macro-level planning/policy instruments.

Gender mainstreaming again emerged in conceptual importance in the second UN Decade within the *Nairobi Forward-Looking Strategies* (1985), which urged governments to address women's poverty and unemployment in their national planning at a "high level." The *Forward-Looking Strategies* observed that

> Governments should seek to involve and integrate women in all phases of the planning, delivery and evaluation of multisectoral programmes that eliminate discrimination against women, provide required supportive services and emphasize income generation. (UN Economic Commission for Latin America and the Caribbean CDCC 15.03.00, 3 1986)

Post-Beijing, gender mainstreaming received further endorsement and conceptual clarity from the UN Economic and Social Council (ECOSOC), which defined mainstreaming as

> The process of assessing the implications for women and men of any planned action, including legislation, policies or programmes, in any area and at all levels. It is a strategy for making women's as well as men's concerns and experiences an integral dimension in the design, implementation, monitoring and evaluation of polices and programmes in all political, economic and societal spheres so that women and men benefit equally and inequality is not perpetuated. The ultimate goal is to achieve gender equality. (1997 A/52/3/Rev. 1 chap. IV, para. 4)

This definition signaled conceptual clarity and informed a degree of institutional self-reflexivity within the UN and international donor organizations and non-governmental organizations (NGOs), resulting in a deliberate attempt to apply gender mainstreaming approaches to their internal management systems and policies akin to what would be required of recipient territories (Moser 2005).

It is ironic, however, that the definition of gender mainstreaming most widely circulated and certainly used to guide mainstreaming practices within the UN so readily lends itself to an additive understanding of gender.[4] This definition, as a result of its simplicity, may very well contribute to the extent to which gender mainstreaming is rhetorically embraced by state-managers, hence, my reference to it as innocuously sustainable. However, such a definition inadequately conveys the conceptual and operational shift that was hoped for, and as I will discuss in Chapter 3, it disseminates a narrow, prescriptive understanding of gender that has sometimes been used to undermine more progressive systemic engagements with gender-based power.

However, feminist academics and development theorists have engaged in acts of theoretical and strategic re-framing with the hope of eliciting some of gender mainstreaming's more radical potential. Jacqui True captures the chameleon-like dimension of gender mainstreaming when she describes it as "an open-ended and potentially transformative project that depends on what feminist scholars, activists and policy makers collectively make of it," and that its transformative possibilities lie in its capacity to challenge the "masculine-as-norm institutional practices in state and global governance" (2003, 368). This, while valid, remains optimistic, since what becomes abundantly clear in the Anglophone Caribbean as elsewhere in the global South is that the equity possibilities depend on the political will that governs the landscape as well as the levels of authority that reside in the agencies and agents responsible for mainstreaming.

Notably, gender mainstreaming in the Caribbean has for the past three decades been in a stranglehold of a Westminster system of government, characterized as it is by authoritarian and masculinized forms of "maximum" leadership.[5] I will repeatedly point to the ways in which this administrative system is often counterproductive to the goals of gender mainstreaming in the Anglophone Caribbean. I am not suggesting that the Westminster system of government is any more of a behemoth than other bureaucratic systems of government. Political scientist Ann-Marie Goetz, for example, notes that mainstreaming gender within the national machinery has generally not done well globally (1995, 14).[6] The Westminster system serves as my recurrent reference point because it is the parliamentary model that predominates in the Commonwealth Caribbean.[7]

Gender mainstreaming places gender-just change at the whim of state forces and desire. As a form of state feminism, gender mainstreaming depends on the state's commitment to institutionalizing the terms of gender equity (Stetson and Mazur 1995). It is for this reason that True argues that it is not whether feminist activists are able to avoid cooptation, but rather, whether feminists can afford to not engage the process of institutionalization. True maintains that mainstreamed institutions, despite their vulnerability and weakness, "provide a platform for change by encouraging new alliances and networking among feminist activists, scholars, and policy-makers inside and outside of government" (2003, 372).

Many feminists share True's optimism that mainstreaming practices and policies provide feminists in the South with a set of practices and organizing strategies that hold government officials accountable for documents ratified and/or adopted at the various UN conferences on women (Dutt 1998; El-Bushra 2000; Ali 2002; Kardam and Acuner 2007). Yet, the lack of enforceability that attends documents such as CEDAW and the Beijing Platform makes it necessary to distinguish *coincidence* from *causality*. In both pre- and post-Beijing, we are more likely to find that local demands and strategies of feminist and Caribbean state-managers alike have *coincided with* the goals and objectives of international gender mainstreaming instruments rather than caused their implementation. This differs markedly from assertions that international gender mainstreaming instruments function in a catalytic manner to create gender-just change within state machinery. It is rare to find instances of gender-equity that have resulted from an axiomatic compliance with the terms of gender mainstreaming-related ratification or adoption. This is certainly the case in the Anglophone Caribbean, where gender mainstreaming instruments have been remarkably ineffectual, precisely because these documents hold neither coercive nor persuasive arguments that would encourage power brokers (e.g., politicians and bureaucrats) to rethink privilege and power.

The lack of political commitment to gender equity is more discernible if we place state responsiveness to a gender mainstreaming mandate under pre- and post-Beijing scrutiny. Sonia Alvarez's discussion of pre-Beijing feminist organizing in Latin America, for example, shows a greater degree of compliance and willingness by state agencies to participate in discussions around gender justice in the lead-up to Beijing. A number of extrinsic motivations, such as the provision of funds and technical expertise by agencies such as the United Nations Development Fund for Women (UNIFEM) and United Nations Economic Commission for Latin America and the Caribbean (UNECLAC), characterized the pre-Beijing period. This support proved invaluable in coaxing and cajoling state responsiveness to gender mainstreaming to facilitate the pre-conference preparation.

In addition to these extrinsic motivators, I want to add a very powerful intrinsic motivation: the ruse of modernity. Beijing undoubtedly provided state-managers in the global South with an opportunity to paint a picture of their own modernity in relation to gender equity. However, I use the ruse of modernity to draw critical attention to the ways in which a number of contemporary development indices now rank a nation's level of development by drawing on the state's treatment of its female citizenry. Indices such as the Gender-related Development Index (GDI) and Gender Empowerment Measure (GEM) are, of course, welcome additions to the development landscape. Yet, I examine the ways in which these indices, by their association with Western notions of modernity, rather than questions of ontological value, may lead to an instrumental deployment of equity rhetoric by regional bureaucrats and politicians.

Building again on the distinction between coincidence and causality, where donor agencies have intervened to facilitate and support gender-just change on the ground, this tends to occur at the level of civil society (e.g., feminist NGOs) as opposed to national machinery. Such interventions may sometimes bring their own institutional mandate thereby supplanting local goals and objectives.[8] There are a number of instructive studies on this point. I have my own recollections of being a university-based representative at a Canadian International Development Agency meeting in which the Beijing Platform goals to be funded were so precisely (and rigidly) named that none of the ongoing feminist work discussed at the meeting would have qualified, without significant tweaking, for funding. Similarly, Elora Chowdhury's incisive critique of the United Nations Children's Fund's (UNICEF) intervention into the Bangladeshi women's organization Naripokkho's work with acid-burnt survivors speaks to this point. Chowdhury argues that while UNICEF helped to financially stabilize the NGO, Naripokkho's guiding vision and long-term strategies became compromised by a services-driven approach as opposed to the originally conceived empowerment approach (2005, 188).

TRAVELLING MILES: LOCAL AND GLOBAL INTERFACES OF GENDER EQUITY

To describe interventions such as those discussed above as willfully hegemonic, which may in some cases be true, merely offers a descriptive and does not help us understand the fraught and complicated relationship that women's organizations in the South have, as they struggle toward a more equitable landscape, with intergovernmental organizations (IGOs), bureaucratic structures of government, and donor agencies in the North. For example, Amrita Basu notes that it is misguided to see the transnational women's movement as only a conduit of Northern domination. Citing organizations such as Development Alternatives with Women for a New Era (DAWN), she observes that some feminist NGOs have been able to accept international funding and still maintain their own agenda (2000). In the spirit of Basu's caution, I find it more analytically useful to examine the differentiated effects that occur at the local level once feminist activists attempt to operationalize policies and regulations that have been determined supra-nationally. Turning to the local foregrounds localized interpretations of gender mainstreaming, contextual constraints, disparities of procedure and access, and cultural possibilities and translations. Turning to the local also increases the urgency to historicize gender relations and, by extension, the multiple meanings of gender at work in various institutional and national contexts.

My analytical engagement with the local should not be taken as an endorsement of the local over the global. Indeed, throughout my work, I hope to show how these two spheres are deeply and always-already implicated in each other. Basu, for example, identifies the local as "indigenous and regional and global refer[s] to the transnational" (2000, 69). Her preference for these terms corresponds with the levels at which a great deal of women's activism is organized, namely, within community and transnationally (ibid.). This definition of the local follows others that have emphasized geographic and political demarcations. Basu's demarcations are important ones as they also resonate with the legally recognized spaces of territory, sovereignty, and civil society.

However, in my conceptualization of the "local," I also want to consider the idea of embodiment. Geographically and politically defined notions of the local tend to render the local as static and confined. Within transnational discourses, distinctions between the local and global also tend to invoke tropes between underdeveloped "natives" and developed "citizens." I am interested in finding a way of disrupting these tropes as well as of offering a way of thinking about the local as phenomena that travel multi-directionally, through diasporic and migratory shifts, through embodied turns and flights—a theme I return to in Chapter 6. Further, I want us to envision how such travel may impact equity-driven arguments and advocacy within the context of the state (as a particular configuration of the local). There is no transversal law of success within gender mainstreaming processes; as such, the local becomes supremely important to understanding how mainstreaming's generally derived principles fare at the level of implementation. What I offer here is a careful mapping of what is possible with small-island states.

I use the idea of *embodied spatialities*, therefore, to refer to the ways in which the politics and place within geopolitically demarcated renderings of "local" are simultaneously conveyed through and influenced by embodied migratory shifts and diasporic expansions. Being careful here to not conflate this understanding with the "personal," I reach for an expression of the local that is differentially expressed diasporically (spatially) and conveyed through a sense of belonging to a personal collectivity (embodied). In this way, my use of embodied spatialities is not merely about the body, although for much of the advocacy using this term it does rest on battles *about* the body (e.g., reproduction and sexuality). Rather, it is a term that I am using to think about the ways in which the diaspora both expands and influences the politics of the "local" through ongoing imaginings of what it means to belong to that local, despite geographic distance.

Localized strategies and modes of interpellation are important because they require us to confront what "gender" means contextually. The local garners an even greater degree of importance because gender mainstreaming has become so dependent on state agents' compliance. Consequently, the limitations and possibilities that feminist practitioners have at their disposal are constrained by the idiosyncrasies of place and politics. For instance, in what ways might clientelism affect the possibilities of "gender mainstreaming" in the region?[9] In what ways might gender advocates within national machineries be personally and politically vulnerable in the majoritarian culture of "first past the post" electoral models or serving at the Prime Minister's pleasure? How do equity strategies such as gender mainstreaming become complicit with the state's imagination of itself as a modern entity? And what are the local understandings at work when the term "gender" is invoked? In other words, how does gender take its meanings from cultural norms and practices, from social relations of gender and race, and from discursive practices that have been informed by colonial histories and various ethnic and racially guided modes of interaction?

In considering these questions, I find it useful to think about Latin American Studies scholar Elizabeth Friedman's use of "transnationalism reversed" (1995). As is my position here, Friedman holds a healthy skepticism toward the impact that transnational gender mainstreaming instruments and organizing have on local contexts. In light of such skepticism, her use of "transnational reversed" highlights "how transnational organizing affects national contexts . . . in ways that may be detrimental as well as productive for national women's movements." Building on Friedman's approach, I set about in *Feminist Advocacy* to study "downward links" to capture the variable effects of transnational organizing in the context of the national. In addition to these downward links, I historicize the local. In other words, an analysis of the local opens an opportunity to assess how historical formations (e.g., colonial antecedents and historically constructed gender relations between and among ethnic groups) also influence the variability of contemporary transnational effects and translations. As I will argue, for example, discursive genealogical effects are certainly at work when we come to assess the possibilities of gender mainstreaming in the context of welfare provision for mothers in the region.

DOES MY SUBJECT AGREE? PARSING THE LANGUAGE OF GENDER EQUITY

Equity strategies such as gender mainstreaming are also shadowed by a conceptual murkiness that adds its own umbra of operational difficulty. For bureaucrats, feminists and policy-makers, operating in the world of gender mainstreaming requires that they develop a high level of competence in very technical language. Some of the expertise requires an understanding of what Alvarez refers to as the "UN bureaucratic logic," which governs much of these processes. This logic imposes its own cycles of grants, agendas for equity, partnerships and, of course, produces a degree of stratification at the local level between those who have this expertise and those who do not. Still, one of the most notable requirements within gender mainstreaming is the need for a completely new literacy and policy vocabulary: pursuing gender equity through processes of gender mainstreaming requires a primer in UN-ese. The conceptual onslaught of "machineries," "instruments," "desk/bureaus," "focal points," and "gender sensitization workshops" not only sets the uninitiated on a steep learning curve, but this terrain might also be experienced as mechanistic and alienatingly disconnected from the language of everyday usage. Beyond the excision from the everyday, I am also interested in the ways in which this language perpetuates a form of disembodied-ness that makes it difficult to see not only agents and recipients but the ways in which some bodies are unable to emerge with any degree of intelligibility within development debates.

International Relations scholar Gemma Carney likens the dissonance between the conceptual and the operational aspects of gender mainstreaming to that of a first-year language class in which the "gender expert" is the only bilingual participant. If the class is to go well, the gender expert must carry the burden of translation, interpretation, and fluency development. Of the "expert," Carney notes "if she is not bilingual, then she may, rightly or wrongly, assume that she is unsuccessful and that because the beginners cannot speak to her in the language she is teaching, conclude that they have not learned anything" (2003, 57). I want to explore Carney's analogy to show that there are deficiencies on both sides of the workshop experience that affect the capacity for translation, interpretation, and application of "gender" as an analytical category.

Carney's use of "bilingual" suggests that it is important for gender experts to be conceptually fluent and, at the very least, functionally literate/fluent in the operational context in which these concepts are to be applied (i.e., the participant's language). However, many "gender experts" function in ways that are more akin to an English-as-a-Second-Language (ESL) instructor's having to contend with the need to develop fluency across a wide variety of languages (i.e., institutional mandates and operational procedures and objectives). As with ESL environments, the Gender-as-a-Second-Language (GSL) classroom too often ignores the student's/participant's first language and insists that the student/participant learn and converse only in the primary language of the instructor. As such, gender sensitization workshops, the primary vehicle of GSL training, may cause participants to experience their training akin to that of a short-term language immersion program. (You may be able to order wine off the menu, but you may not be able to talk to the winemaker about the process of winemaking.) In other words, bureaucrats are being asked to develop gender fluency at multiple cognitive levels (e.g., knowledge, applicative, analytic, and evaluative) simultaneously in a strikingly short period of time while possessing a very rudimentary capacity themselves.

Yet, gender mainstreaming requires coalition building with these very bureaucrats who already inhabit strained working conditions and are asked to learn a "new language," to shift from the linguistic to the conceptual, to apply and analyze this concept to their previously familiar work agenda, and to develop evaluative mechanisms through which they can determine whether they have effectively accomplished over time what they have recently learned in a two-day workshop! And all this must be done in a context in which, at best, they have little vested interest in its success and, at worst, they are overtly hostile to new impositions and challenges to the status quo. Carney's analogy does not quite capture the ways in which gender signifiers carry multiple, contextual, and historicized references within them.

These multiple, varied, and historical signifiers lead to an institutional environment governed by what I refer to as policy heteroglossia. Here, I'm using Bakhtin's discussion of "heteroglossia" to highlight the struggle that exists between official and marginal texts (1982). I explore the effects of policy heteroglossia on gender mainstreaming in a bit more detail in Chapter 3. However, for now, this heteroglossic situation offers us a way of thinking about the diverse and differentially positioned rhetorical policy strands that exist within bureaucratic cultures. It is also a useful way to think about the tensions that exist between and among different feminist and mainstream rhetorical strategies and vocabularies within institutional sites responsible for gender mainstreaming as a strategy of equity. Heteroglossia is, therefore, the rubric that I use to analyze the competing tensions that exist within the institutions which have been given responsibility for mainstreaming processes.

This policy heteroglossia also complicates the gender equity lobby of contemporary feminist practitioners. Not only do these practitioners need to know how the varying rhetorical strategies have evolved over time within the state institutions with which they are engaged, but critically, they need to learn to think strategically in relation to these positions as they are often called upon to do so from positions of vulnerability. Policy heteroglossia is, then, an important reminder that institutional cultures are historically structured and complicated and may be a minefield for the most astute mainstreaming practitioners working toward gender equity.

WHAT IS THIS "GENDER" OF GENDER MAINSTREAMING?

Throughout *Feminist Advocacy*, I explore how "gender" has a number of proclivities that lend it a mercurial quality. Among the elements that contribute to this mercurial quality are the sometimes simultaneous and variable deployment of "gender" as a proxy for modernity; gender as a synonym for the category "woman"; gender as an analysis of power; gender as primarily an additive approach; gender as primary reference point for an institutional critique; and/or gender as performative. Further, when we invoke the category "gender," are we as feminist practitioners inviting a conceptual debate, developing an evaluative instrument, or advancing a policy perspective (Carney 2003)?

Some, when offering a genealogy of "gender," take a linguistic approach in order to highlight the early and more common usage of gender as a grammatical marker; grammar as a point of departure is both inane and instructive. The complexity of contemporary gender debates could not be farther removed from its linguistic antecedents of grammar and syntax. Yet, this point of departure is instructive, for as many authors go on to note, "gender" does not exist in many of the world's languages. This absence, therefore, raises the more important question of whose "gender" dominates *gender* mainstreaming and the terms of *gender* equity. It is noteworthy that the definitions of gender that dominate the development landscape do not originate within discussions held in the global South; they do not center non-heterosexual identities as part of the development environment; nor do they necessarily promote radical expressions of gender for heterosexual women or men. Further, the conceptual debates within feminist academic circles have been significantly more robust and productively messy than those that have been more readily adopted into the operational framework of development policy and planning.

Early feminist advocates of gender saw it as the vehicle through which they could challenge claims of "biology as destiny." Sex was biological; gender was not. Gender was socially constructed and, so, could defy the tyranny imposed through biologically deterministic arguments regarding women's potential. Ann Oakley, for example, argued the distinction as follows: "To be a man or a woman, a boy or a girl, is as much a function of dress, gesture, occupation, social network and personality, as it is of possessing a particular set of genitals" (1972, 158). Similarly, Gayle Rubin, a few years later, began to introduce gender differentiation as anything but benign difference, arguing that gender separation was not about complementarity but suppression of difference. In Rubin's work, gender differences were not analogous to

> day and night, earth and sky, yin and yang, life and death the idea that men and women are two mutually exclusive categories must arise out of something other than a nonexistent "natural" opposition. Far from being an expression of natural differences, exclusive gender identity is the suppression of natural similarities. It requires repression: in men of whatever is the local version of "feminine traits"; in women of the local definition of "masculine traits." (1975, 175)

While reiterating the distinction between biology and social construction, Rubin contributed the additional dimension of gender-based inequity. Gender roles, a term then increasingly prevalent in the literature, was not merely a description of what men and women did; rather, it became one of the primary conduits through which men and women learned their "place" and the traits and attributes deemed desirable and "appropriate." By the 1980s, "gender" as a marker of systemic inequity was firmly established and best encapsulated in Joan Scott's operationalization of gender:

> My definition of gender has two parts and several subsets. They are interrelated but must be analytically distinct. The core of the definition rests on an integral connection between two propositions: gender is a constitutive element of social relations based on perceived differences between the sexes, and gender is a primary way of signifying relationships of power . . . It might be better to say, gender is the primary field in which or by means of which power is articulated. (1988, 42)

Scott's interrogation of "gender" decidedly clinched our understanding of a gender order as foundational to various hegemonic and discriminatory practices. This understanding allowed for an explosion of new vocabulary such as "gender relations" and "gender asymmetries," a vocabulary aimed to highlight the inequity that existed within the innocuous, within the natural order of things, as it were.

While Scott drew on "gender" as a way of marking difference *between* the sexes, Butler in her 1988 discussion of gender performativity brought us full circle by interrogating the presumption that "sex" was distinguishable from "gender." Butler would write

> My suggestion is that the body becomes its gender through a series of acts which are renewed, revised and consolidated through time. From a feminist point of view, one might try to re-conceive the gendered body as the legacy of sedimented acts rather than a predetermined or foreclosed structure, essence of fact . . . Indeed if gender is the cultural significance that the sexed body assumes, and if the significance is co-determined through various acts and their cultural perception, then it would appear from within the terms of culture it is not possible to know sex as distinct from gender. (1988, 523)

These four points present an interesting progression. Recognizing "gender" as distinct from sex; as a social construction; as a social phenomenon with inherent asymmetries; and as clinched by the co-constituting interplay between sex and gender have each radically transformed how we understand human subjectivity. These frameworks offer a layered understanding of how gender identity is consolidated. On the one hand, "gender," rather than being innate, becomes a cultural artifact of sorts, a matter that reflects social norms, values, and ways of being. As such, while we all have some manner of gender expression, the specificity and intelligibility of that expression are profoundly local. Further, gender as a critique of power significantly affects how we identify and name hegemonic practices. As a critique of power, gender functions as a positionality marker as well as a strategy of discrimination.[10] This juxtaposition foregrounds the ways in which gender helps us to make sense of our selves in the world and as a platform on which we strategically take political action while recognizing that "gender" is also an axis along which discriminatory practices occur.

Butler brings a discursive analysis to our understanding of gender. She centralizes the dependence that "gender" and "sex" have on rigidly monitored rules, regulations, and descriptions of *each other* for their appearance of regularity and fixity. What we understand as "sex" is always-already informed by our gender-based practices and rules. Similarly, the knowledge that boy "means" not only male but potentially "blue" reminds us that sex is always at work within our gender norms.

I've presented a relatively uncontested operational development which belies the ferment that attends these debates within the academy.[11] Rather than throwing the reader headlong into these debates, I am concerned with exploring how these various definitions can expand the possibilities of gender mainstreaming if deployed operationally. For example, how might gender as power require us to rethink the operational procedures of the very institutions and organizations vested with the responsibility for mainstreaming the goals of gender equity? How might a discursive genealogy of gender help practitioners better understand the multiple frames of reference at work in any institutional contexts when gender is invoked? If we pursue understandings of sex and gender as co-constituted phenomena, what, then, are the implications for understanding sexuality as equally as integral to a development framework as the process of fairly and sustainably providing potable water?

The definitional complexity intimated here stands in striking contrast to the definitions at work within the UN and other development agencies. I've already pointed to the additive tenor of ECOSOC's engagement with gender, which renders gender as an axiomatic inclusion of women and men. Such an approach obscures the power differentials that exist between and among women and men; as such, it makes gender policy very vulnerable to potentially rechanneling resources to individuals who are already privileged. Similarly, the World Bank's discussion of gender is one that foregrounds the distinction between sex (as biological) and gender (as socially constructed). Theirs is a definition that recognizes that "Gender roles differ from the biological roles of men and women, although they may overlap. For example, women's biological roles in childbearing may extend their gender roles to child rearing, food preparation, and household maintenance."[12] They later speak to the importance of redressing gender disparities within institutional, social, legal, and religious arenas.

While gender disparity primarily manifests as a condition of material inequity, embedded within it are ontological questions about value and worth; matters that always attend questions of inferiority/superiority. Approaches that privilege the material elements of inequity are unquestionably important, but in isolation they can lead to a disproportionate emphasis on an outcomes/efficiency-oriented approach (Sen 2000). The danger here is that this efficiency approach leads to an emphasis on instrumental "problem solving" that is primarily market oriented, often to the detriment of the more complicated yet equally as relevant questions of process and human worth. In addition, such an approach privileges if not prescribes a worker-driven understanding of women's experience of gender and subject formation, thereby bringing gender equity strategies into a very complicit relationship with the tenets of neo-liberalism.

The conceptual and operational chasm between academic and donor agencies is great and not without consequences. Foremost among these is an oversimplified representation of gender within the operational sphere. To this end, feminist practitioners have begun to question the development landscape's reduction of gender's complexity to quick and sometimes inaccurate slogans, many of which are now recognized as integral to some of feminisms' most persuasive rhetoric (Baden and Goetz, 1998; El-Bushra, 2000).[13] For these authors, the tendency toward sloganeering, despite its best intention, also signals the increasing depoliticization of gender mainstreaming. Sally Baden and Anne Marie Goetz, for example, challenge the preponderance of gender sound bites that lend an aura of fundability and respectability to gender mainstreaming (1998, 21).

I am very much in agreement with Judy El-Bushra that "gender" cannot be taken as an "unquestionable good" but am not convinced that we can dispense with gender, if only because of the extent to which it is embedded in how we make sense of our lives and intersecting worlds. If we are to talk beyond the sound bites, then we need to insist on a degree of complexity in our conceptualization of "gender" while maintaining a commitment to the fact that gender cannot be easily disentangled from other aspects of one's identity, a matter that makes coalition building in political work imperative.

EQUITY THROUGH PRACTICES OF COALITION: MAPPING INTERSECTING TERRAINS OF POWER

My final turn here is to suggest that if we are to achieve gender equity in the Anglophone Caribbean, then feminists may need to de-center gender. Heresy this is not—in reality, I am merely suggesting that effective equity work cannot be sustained through a single identity marker. What I hope to make clear in *Feminist Advocacy* is that gender equity also cannot be achieved by treating one's identity as a descriptive process in which we hope that enunciating the identity category (woman, or black, or queer) will result in progressive political effects. Similarly, social justice and equity concerns are not necessarily enhanced by multiple enunciations, that is to say, merely broadening the field of identity markers through the addition of class and race to gender. Admittedly, the act of recognition is political, and the process of acknowledging the existence of people who belong to various ethnicities, genders, and classes is important. However, my point here is that understanding where the line of in/equity has been drawn must be contingent on the categories that become *salient through the field of play in which power reveals itself and through which bodies are created as marginal*. This approach to equity is, therefore, one that is contextual and flexible and acknowledges that power manifests differently on different bodies.

Understanding the contextual, flexible nature of power as well as the ways in which power manifests differently on different bodies encourages us to organize in less prescriptive ways. While discrimination is not random, it can at times be unpredictable. When we begin our analyses with pre-determined categories, it also means that we presume to *know* where our allegiances lie, leaving us with little room for shifting allegiances and coalitions. A commitment to *equity rather than identity* would not only dissuade us from granting axiomatic primacy to race, gender, or nationality, but it would also help to extend our commitments to forms of discrimination that we may not immediately identify as "our own." This is what I refer to as equity through a practice of coalition. Such a practice requires that within our political organizing we think centrally about what is needed by those who are unlike us and differently positioned to ensure that they are able to have similar levels of access and possibilities.

My thinking on equity through practices of coalition is informed by a black and transnational feminist critique. I am sandwiched on the one hand by Bernice Johnson Reagon's reminder that coalitions are discomforting locations. These are spaces, she observes, in which we would prefer to "keel over and die" (1983, 356). However, dying, in so far as it metaphorically represents paralysis, disengagement, isolation, is not an option that Reagon allows because our survival depends on learning how to engage the very spaces that make us uncomfortable. On the other hand, I draw on Chandra Mohanty's call for a politics of solidarity (2003). Hers is a form of border talk, which invokes transnational border crossings, migratory flows, and the need to communicate across historically drawn "divisive boundaries." Further, at the border listening becomes a profoundly political and ethical act.

By no means should we misrecognize my privileging of equity over identity as a dismissal of the political work that can and is done in the name of identity. I am, though, invested in tempering our tenacious hold on an identity to the exclusion of others' subject locations and the organizing that may be required on behalf of and alongside of these subjects. However, envisioning equity through practices of coalition is a theme that I address in the final chapter, which centers queer subjectivities.

I end *Feminist Advocacy* with a discussion of equity as a practice of coalition because it signals a feminism that takes us outside of ourselves to a political engagement that is radically encompassing and transformative. The opening epigraph for this chapter outlines the generative effect that communities of readers offer to the writing process. By imagining an audience, I as writer envision a companion for whom I hope this book will matter. For Flax, imagining a community of readers helps the writer to "constitute outsiders and enemies, who it is necessary and correct to attack." I part with Flax here because my call that we imagine a feminism outside of ourselves makes it difficult, if not treacherous, to too quickly name outsiders and enemies. Equity through a practice of coalition is, of course, not a politics of everything. Rather, it is a politics that signals to state-managers that discrimination against one is inherently laying the ground work for discrimination against all. As such, throughout *Feminist Advocacy* it is my hope that my reader will be prompted to think of new, less exclusionary ways of organizing toward the realization of equity in the region.

2 Crafting Maternal Citizens
Historicizing Institutional Subjectivities within Gender Mainstreaming

> Making visible the experience of a different group exposes the existence of repressive mechanisms, but not their inner workings of logics; we know that difference exists, but we don't understand it as relationally constituted. For that we need to attend to the historical processes. . . .
>
> Joan Scott, "The Evidence of Experience"

INTRODUCTION

It is not enough to merely "see" women. Achieving gender equity must be guided by an understanding of *how* different constituencies of women have come into being over time within their various local contexts. It is now a fairly commonplace understanding that inherent in the pursuit of gender equity is the desire to denaturalize who and what gets marked as different. What is not always immediately obvious is the importance of exploring *how* ideas of the *individual as different* are historically constituted and produced through multiple sites of power. Recognizing which individuals, groups, organizations, and ideas require, or continue to benefit from, women's subordination, queer invisibility, or ethnic destabilization is critical to inequity's undoing. In this chapter, I am arguing that dismantling inequity requires that we learn *how to see* the sites through which specific marginalities have been produced, the tropes that keep these marginalities in place, and where the investments in their perpetuation lie.

Arturo Escobar reminds us that development discourses are replete with their own regimes of representation (1995). This "regime of representation" (ibid.) reflects a world that has been divided by political and economic power-brokers into under-developed and developed territories. This division has simultaneously generated an industry of "problem areas" and "problem solvers," which, unsurprisingly, mirrors historical relations of colonial exploitation. The excision of history from contemporary development debates, as I've stated earlier, sets the frame for poverty to be discussed as a condition rather than as an emergence that results from a set of exploitative conditions. However, one of the consequences of such an excision is that women also feature within development discourses as subjects without history; a phenomenon to which feminists have also contributed.

Ester Boserup's *The Conditions of Agricultural Growth: The Economics of Agrarian Change under Population Pressure* (1965) is now rightly cited as the groundbreaking and transformative work that challenged the gender disparities of development. Her work was pioneering in that it provoked a seismic shift from development as a debate among nation states to the differentials of the sexual division of labor. As an economist with a specialty in agricultural development, Boserup made two significant interventions. First, she found that women were responsible for as much as 80 percent of the agricultural work that undergirded African economies and that much of this labor activity went unrecorded. Second, she made a clear connection between the disparities of the sexual division of labor and systemic control of women in the wider society. Not surprisingly, the trajectory which would follow with the feminist intervention of Women in Development (WID) focused on women's time, labor, and access to resources and the relationship between the public and private spheres of the economy.[1]

Boserup stands premier among an early cadre of feminist thinkers who attempted to grapple with women's marginalization from development policies. Although these feminists were pioneering in helping us to "see" women within development, they taught us to see women somewhat uni-dimensionally. On the one hand, this uni-dimensional perspective presented dependent women in a binary relationship with exploitative men (which often became technology). On the other hand, this pioneering work became reductive through the elision of development with economic progress and women's participation in the market.

This early scholarship insisted that development paradigms such as modernization approaches accounted for women's invisibility yet simultaneously made women in the global South visible primarily through categories such as labor, potable water, and access to infrastructure and other resources. As such, women come into development without history. They are further conceptually flattened out through a global dissemination of "gender" within international equality policies that focus on the social relations that exist between men and women without exploring the history of these social relations.

My methodological move in this chapter calls for a historicization of "gender" within the conceptualization of *gender equity* projects. Historicizing subject formations within development reminds us that bodies are interpellated into discourse, called forth, as it were, to fill a series of different narrative functions within international and national development debates. Because these narrative functions are replete with material effects, I am centering a methodological approach that calls for a historicization of gender within the local. I do so namely to unveil the tropes that hover around marginal bodies, to deconstruct the narratives that are initiated (thereby limiting the available language with which subjects are able to talk about themselves within development debates), to explore the ways governmental institutions and organizations provide homes for these historically constructed understandings of gender, and to point to a form of gender collision that invariably occurs as the "gender" of international policies brushes up against the historically constructed set of social relations within these existing institutional frames.

To this end, I invite the reader to consider *gender-as-genealogy* as a way of reading the historical formation of gender relations and the contextual construction of social relations over time. As an analytical approach, gender-as-genealogy examines the truth narratives that infuse local understandings of gender performance. It presents gender behaviors and identities in their simultaneity as contemporary manifestations and historical constructions. The recognition of gender as historically constructed social relations makes it possible for us to engage the regimes of representation and apparatus (e.g., colonial policy) that consolidate these social relations. By turning to these historical antecedents with critical questions about identity production, we are better able to reposition gender identities that have been foreclosed through the deployment of these regimes and apparatuses. In this sense, I am working with Michel Foucault's positioning of genealogy as that which "allows us to establish a historical knowledge of struggles and to make use of this knowledge tactically today" (1980, 83).

It might be best to begin this exploration of gender-as-genealogy's analytical relevance with a bit of the ironic; namely, that ignoring the historical relations of gender is a strategy that is employed by both "progressives" and "conservatives" alike. Arguably, and certainly for their own strategic purposes, both gender progressives and advocates of a gendered status quo have their own investments in rendering contemporary gender behaviors, practices, and roles as ahistorical. For example, when political leaders declare that none of their citizenship is queer or when Caribbean male marginalization theorists proclaim "so man stay,"[2] what we find are status quo advocates attempting to present a naturalized order of things.

Similarly, we tend to label as progressive those approaches that work toward gender justice—approaches, which, in their attempts to realize a more equitable future, challenge contemporary gender expressions that limit change. However, in pointing the way forward, practitioners at times dismissively relegate the gender expressions that they wish to challenge as a thing of the past. Such a categorical dismissal, I argue, limits practitioners' capacity to understand the embedded nature of these practices, how they are shored up in policy, and, most importantly, how they will re-emerge railing in political speechifying.

Both locations deny the discontinuities and practices that structure the social relations of gender and, when invoked, allow advocates to present a timeless, unchanging social order. The sense of timelessness suggests that subject formation stands outside of history, and as Foucault notes, "enables us to isolate the new against a background of permanence and to transfer its merit to originality.... to the decisions proper to individuals" (1980, 21). Through the incorporation of gender-as-genealogy into mainstreaming approaches, I challenge this desire for gender permanence by reading for the historical and contextual institutionalization of gender.

Historicizing gender becomes important to gender mainstreaming in three very specific ways. First, mainstreaming is essentially a form of advocacy which, in its quest for gender justice, must reckon with what *has been said* about gender. As a strategy of equity, mainstreaming also works to free those subjectivities that have been made invisible by the tyranny of what has not been said. Mainstreaming as a strategy of equity consequently challenges the state-based statements that allow some subjects to emerge as visible, others as abject. Finding these spaces requires a genealogical critique. Second, because discursive antecedents cannot be rhetorically dismissed, by engaging gender-as-genealogy mainstreaming, practitioners are better positioned to craft advocacy strategies that counteract the "inner logic" and ideologies that are at work within institutional locations and seemingly innocuous policy perspectives. Finally, gender-as-genealogy matters primarily because mainstreaming engages with national machinery, and it is here that we find much of our normalizing apparatuses at work. Consequently, as mainstreaming practitioners and agencies are shunted from one ministry to another, we must remain cognizant that there are historically constructed understandings of gender relations already in play within these institutional spaces.

As the epigraph that begins this chapter suggests, the absence of genealogical critique leaves us unable to account for the unities that have consolidated over time within the gender differences that are being challenged by practitioners. We are at best left with the capacity to describe gender relations only as they appear(ed). The effects of discursive practices cannot be rhetorically dismissed simply as a matter of the past; discursive effects will not be wished away, regardless of how undesirable we might think them. Effective feminist advocacy has to consciously engage the discursive formation of gender relations within the institutional frames in which gender mainstreaming will be implemented. By tracing the institutionalization of women within the Trinidad and Tobago state machinery, I show how contemporary advocacy and policy formulation are always-already in dialogue with this discursive interplay between humanity and the maternal.

This chapter is a methodological engagement with the process and effects of reading for gender-as-genealogy within institutional frames. While I attempt to show how and why gender-as-genealogy matters, I do so with the awareness that this historicization, though important to the quotidian practices of mainstreaming practitioners, is not an easy task, nor one I wish to prescribe. I am more concerned with directing the reader toward what we might see from the vantage point of gender-as-genealogy. This is not quite the same as suggesting that such a methodology must always be employed as part of mainstreaming practices of equity. This chapter, therefore, is a reminder that the discursive effects of gender formations are already at work within the institutional cultures in which contemporary mainstreaming practices occur.

I focus my attention on an analysis of the gender formations that attend the provisions of welfare and social services in Trinidad. I center this ministerial location in my discussion because gender mainstreaming agencies in the Caribbean have been disproportionately located within what is traditionally known as "soft" agencies, such as welfare and social services agencies. My objective is to invite some consideration of the tacit gender assumptions already at work within these institutional spaces in which mainstreaming machinery reside. This methodological analysis allows us to discuss how these historically constructed and institutionally coded identities (e.g., the sacrificial mother) affect contemporary gender mainstreaming practices.

MAINSTREAMING ANTECEDENTS: GENDER AS A GENEALOGICAL CRITIQUE

In the plethora of gender definitions that inhabit the mainstreaming landscape, very few, if any, arrive in the South with adequate flexibility for negotiation and engagement with local expressions. Conceptually, "gender" arrives in training manuals, policy approaches, and grant parameters as a pre-packaged category to be implemented through effective gender mainstreaming. These criteria are to a large extent unresponsive to existing formations of gender relations within the local. When local social relations are in alignment with the spirit of advocated definitions of gender, they are lauded. When they are not in alignment, they are targeted for change. Governed as it is by models, frameworks, and definitions, "gender"-related practices seem to offer little more than what Marnia Lazreg has referred to as a "tool of conceptual management"—a tool, she goes on to suggest, that facilitates the imperatives of international capital (2002, 135). Lazreg's connections between gender as a concept and market imperatives squarely places the concept as an organizing axis within the economy of development (e.g., need for experts, manuals, technology, grant writing, insertion of women into global economy through micro-enterprise, and so on (ibid.).

As a "conceptual management tool," the pre-packaged gender modules and manuals impose on the South another entry point for hegemonic thinking regarding the "appropriate" social relations of gender. I am not at all suggesting that the local exists as a pre-discursive site with a superior narrative of gender equity. Nonetheless, it is important to consider how these pre-packaged definitions may silence women and other marginalized constituencies from participating in development debates if their lived experiences of their gendered identities are not part of a more flexible knowledge order of gender within gender mainstreaming. Engaging local gender expressions and vernaculars of the local beyond needing to "fix it" is also important because these vernaculars are already at work, prompting the conceptual gender collision to which I referred earlier.

Ahistoricity fixes gender as a contemporary or futuristic challenge. My contention here is that if practitioners dismiss the genealogical formations of gender within the local, they limit their understanding of how "gender" becomes a kind of conceptual shorthand for discriminatory statements, "the way things are," and we well know that the work of any gender hierarchy is to foreclose possibilities. For gender mainstreaming to effectively engage questions of equity, it must reside in precisely these spaces of foreclosure while engaging (as distinct from merely challenging) how and why people cling to normative understandings of gender even when it may appear to be against their own self-interest.

In an attempt to explore how gender-as-genealogy might work, I look at the over-determined historical representation and production of Afro-Trinidadian women as maternal. I explore how this over-determination is structured over time through state formations. I further examine how this historical antecedent constrains and contextualizes women's contemporary interaction with the provisions of social services.

In the discussion that follows, I want to show gender-as-genealogy's methodological usefulness to contemporary debates on gender equity by examining the gradual consolidation of Afro-Trinidadian women's identity through the frame of the maternal within three watershed periods: 1807–1834, 1930–1962, and (contemporary enactments of the symbolic maternal between) 1990–2001. My periodization does not aim to find some kind of maternal commonality across time. Rather, I aim to explore the politics of consolidation around the black (M)Other, as evidenced by specific mechanisms of government, such as law, colonial responses to social unrest, institutional formations, and contemporary welfare practices. This becomes an interesting way of thinking through the importance of genealogical analyses to mainstreaming in so far as it would allow practitioners to engage the ways in which discursive gender formations affect contemporary mainstreaming strategies or foreclose on those subject locations, which, although equity related, are treated as anathema within Caribbean mainstreaming discourses (e.g., reproductive equity and sexual minority rights).

FROM CHATTEL TO MOTHER: A PLATFORM OF RIGHTS

This discursive examination begins with the supposed contemporary unity of "maternal centrality" which governs political rhetoric within the Anglophone Caribbean. In challenging this unity, I aim to study its internal configuration, as Foucault argues, "just long enough to ask myself what unities they form" (1972, 26). As such, I explore three consolidating turns that shape the way we understand Caribbean women: (M)Otherism, maternal dysfunction, and institutionalized disciplinization. The philosophical premises that guide this discursive examination are that first, when Afro-Caribbean women make the transition from chattel to "human," they do so as mothers in service to the (re)production of the slave population. Second, the growth of welfarism in the Caribbean—and in Trinidad specifically during the 1940s to 1960s—was critical because it eventually led to a policy focus aimed at constraining maternal sexuality. Moreover, this effect was supported by the anthropological and sociological theoretical formulations of the day, which associated the maternal with ideas of dysfunction. Third, these antecedents resonate in contemporary mainstreaming issues regarding appropriate programming for women, reproductive equity debates, and women as sexual agents.

Caribbean feminist Rhoda Reddock makes the argument that due to the legislative turns and gender shifts in labor demographics, the Apprenticeship period (1834–1838)[3] should be seen as the genesis of a housewifization[4] process for Afro-Caribbean women (1994, 32). If this is so, then the period of Amelioration[5] (1807–1834), similarly, because of the legislative responses to women's reproductive capacity, should be seen as the emergence of (M)Otherism in the Caribbean. I am using (M)Otherism to refer to a discursive representation of the maternal as a known "Other," fixed in psychology, practices, and im/possibilities. With this term, I point to the establishment of a particular subjectivity, differently consolidated over time, which emerges in different institutional contexts to serve a range of political purposes, while disseminating a sense of what it means to be a woman.

Gender-as-genealogy philosophically draws on cultural theorist Homi Bhabha's "The Other Question," in which he examines the need for fixity in the colonial imagination of the Other. This imagined other is an identification that "vacillates between what is always 'in place,' already known, and something that must be anxiously repeated. . . . For it is the force of ambivalence that gives the colonial stereotype its currency; ensures its repeatability in changing historical and discursive conjunctures; informs its strategies of individuation and marginalization . . ." (1994, 67). Bhabha outlines a number of processes that feed into the discursive production of subjectivity within colonial and post-colonial[6] national projects. These include the need to produce an "Other" that is both excessive yet credible because the location remains im/possible to contest. I contend that during the Amelioration period these processes consolidated with rabidity on Afro-Caribbean women's bodies and their re/productive capacity.

This Amelioration period allows us to analyze a number of demographic and ontological shifts that occurred between the abolition of the Atlantic slave trade in 1807 and the abolition of slavery in the British Caribbean in 1834. Notably, between 1807 and 1834, the slave population in all of the British West Indies declined from approximately 775,000 to 665,000. Demographic historian Barry Higman attributes this decline of 0.5 percent per annum to subsequent reduction in transported labor following the British Parliament's abolition of the Atlantic slave trade (1995, 72). It is more important for our analysis to question the consequences and repercussions of legislative changes, one of which would have been the colonial planters' increased interest in the question of "natural increase" and the other being the manipulation of slave women's bodies to serve as shock absorbers for declining profits, a manipulation that re-crafted the role reproduction played in Afro-Caribbean women's self-identification in the "New World."

The struggle over the cessation of slavery among the colonial planters, the abolitionists, and the British Parliament signaled a simultaneous struggle for slave women's marked bodies and the symbolic location of the black (M)Other. This power struggle enacted a further colonization of Afro-Caribbean women's bodies because their maternal function became critical to the continuation or demise of British slavery. This battle, however, would potentially be a zero-sum game for the women themselves, in which this identity of (M)Other was arguably spawned out of laws and systems that would secure profit rather than personhood. While the new social stratification firmly mirrored the relations of production, I have argued elsewhere that enslaved women used these profit-driven legislative changes as a platform on which to make claims of personhood.

A significant feature of the Amelioration period (1807–1834) was the heightened attention that abolitionists and colonial planters alike paid to women's reproductive capacities. Planters and abolitionists did not focus merely on the biology of slave women's reproductive capacity. Rather, theirs was an embodied interpellation of slave women's bodies into a battle of ideologies, mercantilist and humanitarian. This shift in legislative focus highlights the presence of a gender differentiation within slave economies that was premised on the production value of the womb. If we conceive of each piece of legislation and (re)presentation of the black woman as an embodied one, in which the female anatomical landscape is being (re)written and narrativized through colonial fictions, what can we learn about the construction of motherhood, sexuality, vulnerability, and female strength and resistance?

Any historical thesis that attempts to divorce British colonialism's racialized systems of production from that of reproduction would have to erase black women's differentiated sexuality, reproductive capacities, and labor as "mothers" to do so convincingly.[7] Examining the interstices of these spheres facilitates a dialogue about how the forces of production chisel and shape female flesh into a body that is first and foremost a functional, instrumental unit of labor from which it extracts physical and physiological labor. Land preparation, field harvest, milling, and punishment (as a further disciplining of the body) feature prominently in the very solid body of Caribbean historiography (Morrissey 1989; Bush 1990). Planters, to enforce the relationship between labor and self, had to create an ideology that only understood the maternal body in terms of its productive capacity. This, of course, does not mean that psychosocial needs did not exist for enslaved women; what it suggests is an ontological impasse that arose within the maternal self due to the coercive nature of the forces of production over the maternal.

Prior to the abolition of the slave trade, though, the relations of production (i.e., slave women as work unit and labor-intensive work) certainly pressed against female flesh such that it would respond more readily to physical productive labor. When colonialists did recognize slave women's maternal identities, the motive was the same: profit maximization. Consequently, little within slave society would permit or encourage any non-productive maternal identity for slave women. More importantly, the ways in which slave women did sometimes express their maternal identity (e.g., infanticide) could not be read within this epistemological framework as an act of mothering since there was no will to understand the complexities, constraints, and frustrations that would make women's pathological performance of their maternal identities perfectly normal in light of the systemic denials.

Against these epistemological erasures, any reproductive act would be muted or filtered by the dominant modes of being and emerge as deficient. While amelioration polices were afoot, there was nonetheless the prevailing ideological battle that placed enslaved women in a double bind of (M)Otherism, always-already produced as dysfunctional, deficient mothers. The idea of the perfect mother informed by the cult of true womanhood was symbolically unattainable for white women and far less so for enslaved women. The prevalent definition of motherhood required self-immolation and encompassed rearing service unto God and Empire. The following excerpt that ran in 1826 in the British-owned *Port of Spain Gazette* gives us an indication of how idealized colonial mothering was configured in the popular imagination:

> If anything in life deserves to be considered as at once the exquisite bliss and preeminent duty of a mother it is this—to watch the dawning disposition and capacity of a favourite child; to discover the earliest buds of thought; to feed with useful truths the inquisitiveness of a young and curious mind; to direct their eyes yet unsullied with the waters of contrition, to a bounteous benefactor, to lift the little hand yet unstained with vice in prayer to their Father who is in Heaven. (September 27, 1826)

By this definition of motherhood, enslaved women by the very terms of their existence as slaves were excluded from attaining the status and rights of mothers. In the 1820s, at the time of this rendition of appropriate mothering, an enslaved woman, if a field hand in a British Caribbean colony, would have worked an average of 3,500 hours.[8] By Higman's calculations, after 1823, the average field hand in Trinidad would have toiled approximately 3,200 hours per annum. Although enslaved mothers were supposedly allowed to turn out two to three hours later to the field, they did work an average of ten hours a day. We can compare this stipulation with the

Trinidad Ordinance of 1800, which had earlier stipulated a workday that began at 5 a.m. and ended at 6 p.m. out of crop season. Enslaved people were given a 30-minute break and two hours at lunch but were still required to feed livestock during lunch and at the end of the day. During the day, old enslaved women who were no longer "productive" cared for children.

Moreover, those who worked the longest hours also worked the heaviest tasks. For enslaved women in the first gang, women were either equal or greater in number than men. This rigorous labor would occur often during the peak of their childbearing years, since the first gang included women between the ages of 18–45. Finally, the day did not end until slave hands "threw grass" (fed livestock) and women tended to household chores. In this context, it becomes apparent that the above call to mothers to discover "earliest buds of thought" was by no means a description that could fairly incorporate the idea of mother as enslaved.

The definition of "Ideal Mother" commonly wielded by the British colonial in the early nineteenth century constrained slave women from acceding to the category of the "good" mother. This onerous and harsh system, consequently, produced a series of pathologies, which when read against the dominant British colonial model of mothering, further ensured that black mothers would always be "Other Mothers." The incompleteness of the black (M)Other was constructed via a number of sites and entered into the realm of truth making through the foundational pillars of colonial law and science and, as such, circulated with the ease and rapidity of rumor and gossip.

Amelioration legislation, by highlighting slave women's reproductive capacity, simultaneously increased the importance of mother-centered family within the productive imperatives of the plantation system. Nineteenth-century colonial forces instituted a profit driven pro-natalism that would in the twentieth century find new life through the anthropological taxonomy of "matrifocality." In other words, what anthropologists would later label as "matrifocal" emerged in no small part from the emphasis that colonial planters placed on motherhood in order to perpetuate the terms and terrain of slavery.[9] (M)otherhood for women and their accession to such became a mark of (B)eing in tandem with, rather than resistive to, the "humanizing" thrust and rhetoric of amelioration. The historicized body of black women became a shifting cultural and economic landscape on which the previous invisibility of maternal physiology was now plotted as a critical component of women's potential humanity.

A number of incentives were given to slave women to support and encourage "natural increase" as one of the defining characteristics of Amelioration. Slave women with varying consistency were to be rewarded for childbearing. To this end, different pieces of legislation were passed; some extended rewards to owners and/or overseers for increased breeding patterns, and others aimed to improve the material conditions of mothers who had given birth. From as early as 1792, Jamaican slave law stipulated that owners encourage their slaves to "breed" by paying three dollars for every slave birth, a sum that had to be divided among mother, midwife, and nurse. Similarly, the

Trinidad Ordinance of 1800 regulated the hours of work for pregnant and nursing mothers, and, although it remains unclear about prenatal care, it stipulated that women who had given birth were not required to return to the field until perfectly recovered (Higman 1995, 350). Lest we be misguided into thinking that these policies held any concern for an emerging maternal subjectivity, we need only look at the 1817 proposition in Trinidad to award silver and gold medals to owners and overseers as incentives for (re)productive slaves. These incentives arguably promoted a number of infrastructural developments to facilitate the increased attention to the growth of (M)Otherism. This included lying-in houses and changes to the hours and quantity of work given to women professing to be pregnant, while simultaneously rewarding owners as one would for prized horses.

In this second ameliorative phase of slavery, "natural increase" was the planters' best defense against the growing movement of abolitionists in Britain. The term "natural increase" is riddled with ironies because there was very little that was "natural" about the marked incentive process required to encourage a reproduction of labor force. The absence of reproductive increase among slaves strengthened the abolitionists' argument that women's compromised fertility could only be as a result of the rigors of slavery. For example, in 1786 a number of Barbadian absentee planters systematically promoted a regimen of "good treatment" for female slaves as a means of justifying the slogan they intended to rail against leading abolitionist William Wilberforce. They argued that "if Negroes are fed plentifully, worked moderately and treated kindly, they will increase in most places; they decrease in no place: *THE INCREASE IS THE ONLY TESTIMONY OF THE CARE WITH WHICH THEY ARE TREATED"* (Beckles 1989, 98; Higman ibid., 304, italics mine).

The pro-natalist sentiment expressed by planters during Amelioration highlights the ways in which colonial mores and laws produced a black female body as a text on which desire, economy, and profit were being inscribed. Additionally, this variability captures the ways in which subjectivities, particularly black women's subjectivities, are produced through a layering of fiction upon fiction. By this, I mean that what was primarily a desire to create profit could only be accomplished through the reclamation of the enslaved female's body. This body became a political pawn in the intensifying battle between slave owner and abolitionist. Black women were to produce, in all the multiple meanings of the word "produce": profit, offspring, and an adequate counterargument for the abolitionists. Yet, in doing so, they would further seal their servitude, as the intent and interest in the production of a black maternal identity, by both abolitionists and planters alike, was never designed to empower black women themselves.

The Order of 1824, as part of a general Amelioration policy, appointed a Protector and Guardian of Slaves. This Ordinance gave slaves the right to purchase their own freedom and choose their own spouse. Slaves, with consideration, could give evidence in criminal proceedings, too. The Order stipulated the number of strokes that could be administered to male slaves. This clause in the Order of 1824 improved on the 1823 Order that explicitly prohibited British slave owners from whipping females, except those under ten years of age.

If we read the archives differently, we will find evidence to suggest that slave women were very aware of these pro-natalist stipulations and strategically positioned themselves to claim these rights. The British parliament designed the office of Protector of Slaves to forestall the end of slavery. Yet, women manipulated the very pro-natal legislation as a means of resisting encroachment in their sphere of operation to develop a platform of rights that allowed them to assert their humanity. The complaints brought by slave women before the Protector of Slaves in Trinidad make this glaringly clear.

By focusing our attention on the kinds of complaints that slave women brought before the Protector of Slaves, we find evidence of women turning the legislation in on itself. Among the array of complaints that enslaved women brought against their owners, we find demands for requisite increases in their allocation of food and clothing (*Port of Spain Gazette* December 20, 1826, page unclear), resistance to beatings that, interestingly, they deemed "unfair" or "unwarranted," or very vocal insistence that a reduced workload was necessary or else a miscarriage would surely ensue (*Port of Spain Gazette* September 9, 1826, ibid). In each case, there was a knowledge of the Amelioration stipulations, a willingness to claim these as a right, and in the process of doing so, an assertion of their humanity through the framework of the maternal as they manipulated the very policy deliberately designed to keep them enslaved for purposes of profit.

The persistence today, a century and a half later, of the symbolic black matriarch/strong black woman suggests that it fills a psychological need to invent a desirable historical "mother" that is necessary for an imaginative return. In short, the historicization of black (M)Othering helps us as black women, the post-colonial state, and former colonial masters to make sense of the horrors of slavery by questioning the validation and production of this symbolic location. This imaginative return, however, cannot be an acritical acclamation of the strong black matriarch without recognizing the complexities of this positionality. Moreover, we must be aware that by readily embracing this historical construct in the frame of contemporary policy, we, too, become complicit in the construction of an overwhelming myth created by the colonial regime for its own devices and economic interests. Women have always had very complicated relationships with their maternal roles, as is immediately evident if we place slave women's corporeal realities at the center of our analyses.

LABOR RIOTS, ACTIVISTS, AND CIVIL UNREST: THE MATERNAL TH(D)READ

Although Amelioration bequeathed a degree of (M)Otherism to contemporary thinking on female subjectivity, the consolidation in the 1930s and 1940s of this representation of gendered womanhood was critical for its scapegoating of mothers for social crises through narratives and the institutionalization of dysfunction within the early welfare services established by the Colonial Office. I frame my investigation in the context of Trinidad's 1930s because the period between 1930 and 1940 represents a notable turning point in the British Caribbean's attempt to reject and question its political and economic ties with Britain.[10] Among the defining features of this decade are the effects of a global depression, the resulting labor unrest of 1937, the arrival of the West India Royal Commission (WIRC) to investigate the factors that led to widespread social unrest in the region, and an impending world war. This particular historical period of political and social upheaval stands as a productive moment of engagement because we can begin to see the emergence of post-colonial concerns regarding the conditions associated with the historical colonial submergence and erasure of the various categories of women and men as the nascent nation experienced significant changes in the nature of its colonial relationship with Britain.

My intent here is not to trace linear time but to strategically examine the appropriation and dispersion of the symbolic maternal as a way of looking for both continuities and discontinuities of the maternal image. My ultimate goal in this gendered genealogy is to later highlight the discursive residual effects on contemporary discussions of equity. This gender travel, therefore, helps us to see how the female body emerges in these political conversations, while noting later how these specters reappear within contemporary institutional frames. As such, it becomes an examination of *negotiated continuity* that requires us to pay attention to the strategies, the calculations, the institutional regulations, and the developmental praxis of crisis management that occur subterraneously and overtly around maternal references. The term "negotiated continuity" points to the repetition of the colonial stereotype of the dysfunctional black mother. It provides continuity in its repetition of dysfunction; it is, though, negotiated in its manifestation and the strategies used to keep this dysfunction in place, as it were, shifting from profit maximization during the Amelioration period to the maintenance of social order and equilibrium in the 1930s–1950s. It produces an overwhelming maternal image that cannot be proved or disproved. Therefore, the likely *probability* of dysfunction legitimizes all attempts to map further social crises on the black maternal body. This act, in turn, limits the effect of individuation, making it difficult for women in their individual capacities to assert alternative narratives of being.

In Trinidad, the years leading to the major riots of 1937 were characterized by a post-World War I increase in the cost of living, the Great Depression, and the simultaneous deterioration of local working conditions. Soaring profits by the oil company Apex Oilfields (Trinidad) Ltd. and the levels of mistrust between employers and employees also proved catalytic. Ideologically, the influence of the Pan African Movement and the sense of outrage and solidarity at the Italian invasion of Abyssinia both contributed to a consolidation of black consciousness and pride.[11] On June

18, 1937, the oil field workers in Trinidad undertook a series of protests led by the Grenadian-born Uriah "Buzz" Butler, who was later joined by barrister Adrian Cola Rienzi. The powerful lobby carried by trade union leaders regionally led to the appointment of a Commission of Inquiry by the Colonial Commission of island-specific commissions into the disturbances. In Trinidad, there was the establishment of the Foster Commission whose findings, published in 1938, recommended the establishment of "good" (read non-militant) trade unions in order to induce and support a more conciliatory resolution to labor disputes (Singh 1987, 276).[12] Despite these individual commissions of inquiry, requests were laid in the British Parliament for a general WIRC to investigate the socio-economic causes of the labor riots.

In response to the labor riots of the 1930s, the Colonial Office attempted to constitute the WIRC with members who had some understanding of the region. Regardless of the colonial agenda driving the work of the Commission, there was a groundswell of support and anticipation by Caribbean citizens for the proceedings, with the hope that it would realize improved living conditions and accountability.[13] The Trinidad Workingmen Association, in a supplement to the Moyne Commission, acknowledged that "The West Indies . . . expect more from this Commission than any previously appointed. Expectations run high. Democratic questions have never been so frequently and earnestly asked" (Memorandum Supplement 872, 72). This enthusiasm was supported further by the rigorous interrogation of government officials to expose what the Trinidad Workingmen Association referred to as "the nakedness of their equivocal and paradoxical position and attitude" (ibid.). As such, a myriad of associations, organizations, and groupings brought their documented grievances before the Commission.

The predominantly middle-class Coterie of Social Workers, itself guided by a self-help mandate, was one of the many organizations which not only submitted memoranda before the commission but also had an expressed concern for the needs of indigent, young black mothers in the 1930s. The line of questioning pursued by the members of the WIRC, as well as the concerns raised by associations such as the Coterie, located the question of mothering firmly on the Colonial and Crown Colony government as a policy agenda while developing a language for articulating the question of maternal rights and expectations of the state ("Royal Commission Hearings," *Trinidad Guardian* February 26, 1939, 12).

The self-help impetus that drove many of the middle-class women's organizations in the 1930s catapulted the question of maternal responsibility into the realm of social responsibility and obligation to and by civil society. These organizations focused on the moral uplift and welfare of mothers. Their primary intent was by no means revolutionary in the contemporary understanding of empowerment and challenge to the discriminatory status quo. However, we cannot dismiss their goal of increasing the income earning capacity of women who had fallen on reduced circumstances, and, in this regard, the Coterie, for example, extended beyond the religious fervor of "charity and good works." During this period, the dominant scripts of black maternal subjectivity oscillated between individuals in need of uplift and blame for larger social instability—both laced with deficiency.

The *Trinidad Guardian* carried a daily report of the Tribunal sitting held at the San Fernando Town Hall to arbitrate the oil field workers' strike. On January 1, 1939, Mr. Till[14], a representative of Apex Oilfields (Trinidad) Ltd., was questioned by the investigating authorities about the competence of the Trinidadian labor force compared with their English counterparts. In assessing the competence of local workers, however, there is an ironic reappearance of the female body in terms that are familiar and in keeping with the tone of maternal dysfunctionality that pervades colonial narratives:

> *Mr. Nelson (Tribunal member):* I take it that your experience is that they are slow in learning and slow in working?
> *Mr. Till:* Very much so.
> *Mr. Nelson:* Of course, you realise in bricklaying there are many men like that?
> *Mr. Till:* There are.
> *The Chairman:* It is not a difficulty due to natural inability or incapacity?
> *Mr. Till:* No, perhaps I could put it that it is due to a great extent to a lack of training.
> *The Chairman:* Yes, and that is a responsibility to some extent of the Government and to some of the employers, is it not?
> *Mr. Till:* It is not the responsibility of the employers in Trinidad, because competent instructors are not available in the island.
> *Mr. Jagger:* I understand you took it further than that. I gather that your complaint was that the mothers did not look after the boys at all properly before they went to school and that the teachers did not teach them anything when they went to school
> *Mr. Till:* That is so.
> *Mr. Jagger:* I gather what you were saying was an indictment both of the mother and the teachers?
> *Mr. Till:* Yes.
> *The Chairman:* The mothers do not look after the technical training of the boys?
> *Mr. Till:* Not the technical training—the home training.
> *Mr. Nelson:* What is their general outlook?
> *Mr. Till:* The general outlook is all wrong due to training.
> *Mr. Nelson:* Due to the absence of training. (*Trinidad Guardian.* January 1, 1939, 2)

The assumed laborer in the riots and the subsequent investigations is once again historically recorded as a male figure. This erasure occurs despite the fact that women were principal activists and agitators in the labor unrests within the oil-belt. More significant is their re-entrance into the debate, not in a political capacity, but as nurturers, both as mother and teacher. This re-narrativization is premised on inefficiency and inadequacy, that is to say, an inability to produce an appropriately modernized worker, prepared with the "machine outlook" and poised to participate in national growth (*Trinidad Guardian*, January 1, 1939, 4).

The excerpt has a strong, if not explicit, subtext of maternal ineptitude. The subtext of maternal culpability begins to function as a truth, consolidated as "a system of ordered procedures for the production, regulation, distribution, circulation, and operation of statements" ready to break into seemingly unrelated narratives (Foucault 1980, 133). This apparent "intrusion" of the symbol draws attention to the ease with which the maternal Other starts to be called upon to justify social crisis or dysfunction. It is interesting that this symbol appears within a completely unrelated discussion, that is to say, a comparative valuation of labor. Yet, interwoven in this discussion is the relationship between maternal nurturing and a supposed poor work ethic. Interestingly, having asserted maternal culpability, the exchange continued onward to a quasi-scientific measurement of the laborer's time, skill, and competence at decarbonizing and grinding valves. This seamless transition naturalizes the causal relationship established between maternal inadequacy and underperforming apprentices, a trope which resurfaced two days later when Mr. Till again establishes that the lack of training in the apprenticeship training period was "due to the period of the boy's life between the cradle and the age of 16 when he comes to us" (*Trinidad Guardian*. January 4, 1939, 2).

My brief, though circuitous, journey through the labor riots of the 1930s and the eventual establishment of the WIRC, better known as the Moyne Commission, offers us an important entry point into the establishment of a development praxis of social-welfare in the Caribbean and subsequent institutionalization of Afro-maternal dysfunction. The social crises of the 1930s were very easily mapped onto an embodied landscape of the black (M)Other and provided a very ready landscape for projecting an increasing imaging of social dysfunction. This was a constant subtext in the proceedings of the local inquiries into the riots and also part of the tenor of the Moyne Commission. Despite extensive participation by women activists and union leaders in the riots, women emerge in the inquiries as inadequate mothers.[15]

THE BIRTH OF THE ANTHROPOLOGICAL EYE/I: DISCIPLINING DYSFUNCTION[16]

The 1950s and 1960s are pivotal to my gender-as-genealogy methodological mapping. While the colonies during the 1930s–1950s pursued and agitated for language that could speak to and identify a national identity, the anthropological thrust of the 1950s and 1960s formulated language to describe the region's familial and social formations; this is language that continues to influence the ways family formations are understood in the region. This period consolidated a taxonomy and gave descriptive language imbued with valences of "natural" and "truth." So, while I have attempted to draw on a historical genealogical analysis to point to the ways in which black women's subject formation was over-determined as maternal in negotiation with the profit making legislative and economic mechanisms of slavery and colonialism, this anthropological taxonomy naturalized ideas of dysfunction and maternal inadequacy.

The dominant structuralist approach toward social engineering envisioned and theorized the family as the barometer of social and political stability. One of the conclusions of the Moyne Commission Report in response to the 1930s labor riots was the social "unacceptability" of the female single-headed family form. The British colonial authors of the Report observed that where there was no father:

> (T)he whole financial responsibility (would) fall on the mother . . . In such circumstances cases of extreme poverty are inevitable, for the standard of living must be lower than it would be in a family group where even if both parents were not employed, more money would be available since the wages of men are normally higher than those given to women. (Moyne Commission 1945, 40)

This excerpt reflected the Report's general tone toward women, mothering, and its role in preserving social order. So, while they accurately acknowledged the disproportional financial burden that fell to mothers, their solution was not to strengthen the structural and institutional framework available to single-headed family forms, but rather, to advocate that girls should be suitably trained and educated "to be companions for husbands." The message of the Report was clear. The stability of the social order depended on the institution of nuclearity, as well as the presence and functioning of the patriarch, husband, and state, all of which were necessary to rein in women's "promiscuity," the related high "illegitimate" birth rates and, last and definitely least, the social fall out of poverty and lowered standards of living.

This link between social stability and family "order" within the Moyne Commission resonated in the anthropological influx of the post-Moyne period. The period of the 1940s–1950s was a tangible historical moment in Caribbean development. One of the first academic responses to the Commission's Report was an attendant influx of anthropology and sociology "experts," ready to diagnose and understand Caribbean aberrations of family patterns. The University College of the West Indies, established in 1946, acknowledged very early in its institutional life that the Caribbean was an unusual and unexplored site of research and scholarship. In their Research Programme and Progress Report (1955, 1), the institution claimed that:

> The West Indies provide abundant opportunity for economic, historical and sociological research in the widest sense. The territories show great variety in their political and racial histories while they sometimes present the same general problems with interesting local differences... We find it difficult to resist the temptation to enlarge upon the intellectual adventures that are waiting in this field. (quoted in Blomstrom and Hettne 1984, 99)

The establishment of a new university, initially staffed by British expatriates, in a geographical landscape that was relatively under-explored, yet inhabited by and hospitable to whiteness due to years of colonization, was just too much to resist. The influx of anthropological and sociological scholars came with the desire to understand a range of fascinating new phenomena, but primarily with the hope of reconciling the weaknesses within the existing family structures so that future social crises could be averted.[17]

This period of social investigation is a critical juncture in the progression of my analysis because of the extent to which the practice of anthropology in the Caribbean succeeded in naming and institutionalizing "maternal subjectivity" as a recurring absence. This discursive performance of anthropology in the region during this period is important because it is here that we find the development of a new literacy, lexicon, and methodology by which to articulate and typify how men and women come to name themselves and, equally as important for my discussion on equity, how these naming practices come to be institutionalized with implications for the terms by which one is able to access state resources. Consequently, the anthropological scientific method constructed and further codified women through the lens of the maternal. This "scientific method" would also eventually provide state actors with the authoritative cloak of "objectivity" and "neutrality" in their conceptualizations of the maternal subject.

This period hides so many of the aspects that we see as the contemporary state's understanding of and relationship to mothers who attempt to lay claim on the state's resources. The anthropological conclusions of the period mark a form of repetition of the naming of Afro-maternal processes and events as dysfunctional. By no means innocuous, these processes, as I have shown, were already historically at work within the colonial antecedent, the 1930s and the 1940s. However, the disciplinary credibility of anthropology, as reflected in its adhered-to jargon, its existing theoretical framework, and scholarly agents to advance its ideas, succeeded in producing another category of seemingly self-evident material/maternal truth.

When we turn to the anthropological impetus that governed the period of the 1950s and 1960s, names such as R.T. Smith, Franklyn Frazier, Melville Herskovits, and Edith Clark figure very prominently. Yet, because I am concerned with the processes by which anthropology helped institutionalize the disciplining of female bodies through a maternal subjectivity, I will focus my analysis on T.S. Simey, a highly respected British sociologist who was also named head of the newly formed Colonial Development and Welfare Office (1941–1946). Simey's very personage was instrumental in forging a disciplinary dialogue among social engineering, gender subjectivity, the family, and academia. The politics of Simey's location, both as a colonial administrator and a trained sociologist, is often ignored within Caribbean sociology literature. However, I maintain that his scholarly/administrative nexus positioned him as a critical embodiment of the linkages that were being forged between state resources and anthropological family studies.

There are several factors which, if reassessed, support this interpretation of Simey's location. First, Simey, a structural functionalist himself, strongly advocated for connections between colonial policy administration, individual identity, and family functionality. He maintained that it was analytically impossible to understand the social conditions in the Caribbean without some attention to the "problem of family organization" (1946, 14). Second, as a sociologist-cum-administrator, he mediated the sociological and political realms in what he had hoped would be a neutral and symbiotic relationship; these are rarely ever so. His hope for a neutral, symbiotic flow between the sociological arguments of the time and colonial welfare planning would be due in no small part because of his disciplinary faith in the scientific method's avoidance of overt value assertions. He viewed the relationship as symbiotic because social scientists entering the region had at their feet the ideal case study environment in which to apply theoretical perspectives to the "problem of social readjustment" in light of existing civil unrest and, therefore, held the view that the Caribbean provided colonial administrators with the opportunity to think dispassionately and in "scientific terms about their daily work" (ibid., 29).

Simey critiqued the areas where the WIRC failed to initiate a stronger tie between administration and "scientific" reform of welfare and planning.[18] His thinking reflected the functionalist vision of his era, and, as such, he argued that the responsibility of a centralized colonial administration was to facilitate a smooth relationship between communities and governmental agencies in the provision of social welfare. The welfare planning model, then, that has become part of our contemporary understanding of planning and policy formulation *cannot* be understood outside of the influence and sentiments that would have been derived, first, from the WIRC and, second, from the functionalist anthropological debates that circulated around its initial formation.

Within the traditional or classical reading of structural functionalism, adherence to and performance of normatively regulated roles with assigned social obligations and expectations are critical to the suspension of social chaos. This significance given to role performance within structuralism crystallizes why the anthropological lens focused so heavily on the maternal, despite the fact that its primary unit of analysis was the family. The notion of the family was premised on an assumed relationship between role performance of individuals within that family structure and the integrity of the social collective. Any attempt at interpreting the apparent dysfunction of the family formations (structural phenomenon) found in the Caribbean required an assessment of the efficiency with which various roles (in this case, the maternal) were executed. I want to argue, however, that in the Caribbean this classical anthropological investigation became remarkably overdeterministic, producing a seemingly one-to-one correlation between the familial/social dysfunction and female subjectivity through maternal narratives.

If we impose a feminist reading on these structuralist approaches and their impact on welfare planning in the Caribbean, a different range of im/possibilities are revealed. Despite the analytical focus on the functionality and organizational structure of the family, the centrality of the mother figure assumes paramount importance in all of the anthropological approaches of this period. This phenomenon emerges from the anthropological appeal to structuralism and normativity within the conceptual approaches that forced Afro-Caribbean mothers to perform a disappearing act from the very investigations that placed focus on them. The supposed universality of Western familial normativity and the expected performance of designated roles on which this normativity is argued help us to understand why investigations into the institution/structure of the family, in turn, became a disembodied typography of Afro-Caribbean maternal identification.[19]

It is against this backdrop that Simey established a typology of Caribbean family forms. In his typology, he distinguished among the "Christian Family," characterized by legal marriage and a patriarchal orientation; "Faithful Concubinage," also patriarchal in its formation, lasting beyond three years but without legal status; "Compassionate Family," which referred to a household formation where individuals aligned themselves with each other because of the convenience; and the "Disintegrated Family," which was essentially a female-headed household. To his credit, Simey's focus on maternal centrality did not extend to incorporate a thesis of maternal dominance. Symbolic centrality, he argued, did not necessarily impute social centrality, and, so, it would be invalid to make an argument that Caribbean women were dominant in their social formation. Consequently, he explicitly repudiated the use of the term matriarchal. Simey argued that Caribbean women were by no means equal and, while they may be seen as matriarchal in function, they remained patriarchal in status (1946, 44). Although refuting matriarchal status, Simey did, however, endorse sociologist E. Franklin Frazier's work in the United States, which prompted him to similarly isolate the "maternal family" as a diasporic phenomenon of post-slave society (ibid., 88) and, thus, a feature of black family formations. Simey's endorsement of Frazier's work marked another phase in the growth of the symbolic black (M)Other. By endorsing Frazier, Simey introduced a transcultural and diasporic line of questioning around the treatment of the black female body and manipulation of her maternal identity in disparate post-slave societies.

Both Simey and Frazier saw the "maternal family" as a consequence of the impact that slavery's economic and social marginalization had on black family formations. These transnational conversations on the black maternal body raised similar questions and conclusions that could lead one to believe that black women were essentially dysfunctional mothers. If we were to trace the impulses that generated similar narratives of black maternal pathologies in the United States, though, we would find similar strategies, performances, and phases of naming of the symbolic black (M)Other (Roberts 1997).

Simey's voice remains an important one to pay attention to because of his role in the Colonial Development and Welfare Office. It is easier to understand why women's lived realities are so absent today from policy decisions when we examine the rationale that informed early thinking on family-related policy formulation—Simey's, for example, placing the need for collective coherence at the foreground of his research. In this schema, maternal realities dissipate and are only factored into policy if attention to maternal role performance abates any threat to collective formation. Indeed, Simey discouraged any pull toward individualized needs, identifying chronic individualism to be one of the sources of sociopolitical discontent in the Caribbean (1946, 24).

I am not suggesting a one-to-one causal relationship between an individual administrator and the institutional conceptualization of the maternal subject. Indeed, the maternal phantasm by this time functioned discursively not simply as traceable moments but rather as excesses of these traces. The body of anthropological work that I am discussing points back to an *amalgamation* of cultural, historical, and popular representations of the maternal subject. Yet, Simey's scholarly location as a colonial administrator cannot be ignored as one of the modus operandi of *institutionalizing* the maternal subject. His influence in the development of the Colonial Development and Welfare Office must be highlighted as a defining moment in which the maternal is deliberately factored into a policy framework but taken by itself does not and cannot articulate all that is invested in this symbolic reference.

The 1950s signaled an ever-expanding network of anthropological and sociological conversations about the maternal body. While Simey's ideas are critical because of his administrative role in the development of a praxis of social welfare, R.T. Smith, to whom I will turn briefly, is critical because, as good language is wont to do, he offers terminology that seemed to distill an essence, a maternal essence—"matrifocality."

The term "matrifocality" figures as one of the most recurrent concepts in studies of the Caribbean family structures. Yet, what is striking is that this term which has come to stand in for Afro-Caribbean women's maternal identities exists in its primary formulation completely without their voices and remains conceptually and operationally uncertain in its own right.

Smith's definition of matrifocality is surprisingly ambivalent. He highlights a family form that was not necessarily linked to co-residential status but had the *legal* and *moral* expectations of co-residential marriage and an increasing salience of women in their role as mothers:

> Thus, the term "matrifocal family" referred NOT to a system of female-headed families, nor to a matriarchal family system, but to a *social process* (emphasis mine), in which there was a salience of women—in their role as mothers within the domestic domain, correlated directly with the class position of the population involved, and focusing on the articulation of kinship and class. (1988, 8)

Here, the term "matrifocal" is deliberately distinguished from any particular family form and does not exist co-terminously with "female-headed households" (FHH).

Consequently, in Smith's definition of matrifocality, the matrifocal family form was possible in situations in which a male was resident in the household. It is the mother's increasing importance to the members of the household that is given pre-eminence. Yet, if this rather generic, all-encompassing net is cast, what, then, distinguishes the "matrifocal" family? In 1956, in an essay entitled "Hypotheses and the Problem of Explanation," Smith's designation of matrifocality initially emphasized the centrality of the mother figure. The corollary of this centrality is therefore the presumed marginalization of men in their paternal capacity even when resident in the household. Smith asserted that within a matriarchal family system, men are marginal even when present within the household. This produced an internal tension where mothers functioned as *de facto* leaders and fathers as *de jure* heads (1996, 13).

It is not surprising that, based on this definition, the label "matrifocal" has been associated with antagonistic gendered relations of emasculated, excluded males and domineering, belligerent females. Smith described the male marginality that results from matrifocality as that of a father who systemically "associates relatively infrequently with the other members of the group and is on the fringe of the affective ties that bind the group together." Smith documents paternal manhood as a commodity that functions in primarily economic, instrumental terms:

> For any particular woman with children, the problem is to find a male in one of the above statuses (i.e., economically viable) to provide the necessary support . . . In the lowest status group the only basis for male authority in the household unit is the husband-father's contribution to the economic foundation of the group, and where there is both insecurity in jobs where males are concerned, and opportunities for women to engage in money-making activities, including farming, then there is likely to develop a situation where men's roles are structurally marginal in the complex of domestic relations. Concomitantly, the status

of women as mothers is enhanced and the natural importance of the mother is left unimpeded. (1996, 16–17)

Smith commodifies the maternal/paternal nexus. Male relevance is, then, premised on financial provision and, further, this economic function is diminished (hence, the increasing maternal salience) as the mother's financial dependence is taken over by her own activities or that of her children. This definition of male "marginality" does not take cognizance of the ways in which *dimensions* of Caribbean masculinity and masculine authority is "ideally" evidenced not by an overt involvement in domesticity, but rather, by a form of looming, if not disciplinary reticence. Second, it is premised on a crass commodification of family ties without the cultural recognition that male financial responsibility is not always seen, especially among already economically deprived communities where Smith based his initial work, as an automatic responsibility and is often undertaken as *an expression of* affective ties (Chevannes 2001). Here, Smith's work is explicitly bereft of any analysis of the internal dynamics of family formation or negotiated gender politics within the family in the construction of masculinities and femininities.

Further, what are the social processes that lead to maternal centrality as a peculiarity that is Caribbean specific? Building on his 1956 essay, Smith in 1973 again revisited his notion of matrifocality. In his later work, Smith advocated for what he referred to as a "matrifocal complex." This complex pointed to an expanded variation of "matrifocal formations" where he focused on structural formations across a wide range of formations and cultural locales, in contrast with his earlier emphasis on a predominantly mother/child relationship as the elementary unit of the family (1996, 54). In this revised schema, his most generalized use of the term was found in a type of matrifocality based on "Domestic Relations." Here, he emphasized the increasing sex-role differentiation that produced a sense of male marginality from domesticity and the countervailing dominance of mothers. As such, the focus was not on the mother/child intimacy but on role allocation where women in their role as mothers monopolized domestic relations. This conceptualization was sufficiently general to be applied in societies and cultural formations as diverse as West Africa, Java, China, India, and the Caribbean. Indeed, based on these references, the greater non-white world could be considered along a continuum of matrifocal domestic relations (1988, 182; 1996, 54).

He premised his second classification of matrifocality on the structure of familial relations; that is to say, preference was given to family networks and kinship over conjugal relations (1996, 55). Among the working class, Smith argued that marriage as a public ritual reflects status and prestige rather than intense romantic involvement. Conjugal ties are agreed on within the working class when it is economically viable to do so rather than as a response to any amorous sentiment. Within Smith's model, conjugal formations bring little to the kinship system, and such long-lasting ties are seen as those that result between mother and child. In a later article, "Hierarchy and the Dual Marriage System in the West Indian Society" (1987), Smith traced the relationship between conjugal activity and class status. He concluded this article by returning to his original argument that economic circumstances dictate whether or not legal marriage is seen as a viable outcome or not. However, we are not told *why* these economic circumstances emerge with such value, nor do we understand any more clearly why the relationship between the child and the woman who "grew" him or her remains the lasting tie.

His final contributing category regarding matrifocality is that of "Stratification"; this he defined as a mother-centered formation that resulted from conditions of poverty, racism, and differential social status. Smith noted that while a relationship may exist between poverty and matrifocal family forms, matrifocality nonetheless existed among a wide range of economic settings. In other words, poverty could produce matrifocality, but matrifocality did not automatically imply poverty.

The problem that we as readers of Smith's work experience is that it provides little or no conceptual stability. Matrifocality becomes all things with all possibilities and, for itself, remains uncertainly not anything at all. Matrifocal families are at once poor and not poor, peculiar to the Caribbean, but found in West Africa, Java, China, and India. Its formation is premised on close ties between mother and child but can be found in a broader household and in expanded kinship networks. It is not to be equated with FHHs but is characterized by maternal centrality and speaks to an increasing level of male marginality over time. Not surprisingly, therefore, Smith concludes his 1973 essay by asserting:

> The coining of terms does not solve problems and may in fact obscure more than clarify. The concept of matrifocal family and domestic relation is a difficult one because of the complexity of the factors involved. In this paper I have tried to set out the major dimensions of the problem as I see them, rather than attempting a "definition" of matrifocality. (1996, 56)

The danger that I have been working toward in my overall critique of the various anthropological approaches is that, despite their remarkably nebulous content, they all coined terms or analytical approaches that have now been appropriated as a disciplinary nominal marker of Afro-Trinidadian women's maternal locations and mothering identities. Their conclusions mask the internal differences of these formations and, by extension, the levels of autonomy and subject formation negotiated by women in these varying family forms.

A critical examination of the anthropological treatment of the Caribbean places us squarely within the discursive relationship between identification and identity. In this relationship, the anthropological Eye/I enunciates an authoritative pull toward *identification* (naming) via a series of disciplinary and discursively structured symbols, systems, and names. Regardless of the (I)dentities that lie buried beneath this signification, the process of naming is so excessive that it produces a female subject that has been robbed of subjectivity; that is to say, void of the capacity to recognize or name self beyond the constellation of anthropological representation. In this frame, to not be maternal is to be in some way pathological; to be maternal is to be potentially dysfunctional. In sum, what emerges discursively and is institutionalized formally is a representation of womanhood that signifies first and foremost as a black maternal subaltern.

I have been focusing on processes of disciplinization and institutionalization in my attention to gender as genealogy because they both mark important moments of codification. Disciplines and processes of institutionalization produce their own codes and procedures of inattention. In many ways, these processes are akin to the production of what Appadurai (1988) calls the "incarcerated native"; here the incarcerated *maternal* is similarly bound by thematic persistence, representational repetition, and symbolic sameness within the anthropological debates. As such, the historical production of blackness and (M)Othering become a mere landscape on which to successfully wor(l)d the Afro-Caribbean maternal reality as *matrifocal* and through processes of institutionalization name the set of goods that it is appropriate for this subject location to desire.

DISCIPLINING THE MATERNAL BODY: PRODUCING DOCILITY THROUGH PANOPTICISM

Tracing gender-as-genealogy enables an examination of the ways in which the (re)appearances of an earlier colonial maternal figuration constrain contemporary questions of equity. I want to now turn our attention to these re-appearances and configurations with particular attention to questions of welfare; first, because it is within these institutional locations that the maternal has been institutionalized, and, second, it is through this institutionalization process that women have had any desire beyond state-sanctioned maternal scripts constrained.

In thinking through notions of equity, it is important to challenge any modes of interaction that privilege an institutional "common sense" about women. As I have just tried to show, this mode is one that is produced within institutions over time and, as I will now show, holds significant consequences for the ways in which women are able to interface with the state.

How are women caught in the nexus of celebration and castigation in which we celebrate the sign of "mother" without commensurate attention to the incapacitating socio-economic components of this identity, thereby dismembering women?[20] How is state machinery implicated in the discourses that have led to a centralization of mothering to Caribbean women's self-conceptualization? How has this identity been manipulated at the state and policy level, often to the detriment of the women themselves?

The discursive analysis that responds to these questions forces us to think beyond the state as disembodied and disinterested and encourages us to see state power as both diffused and mediated through state functionaries, agents, representatives, and managers who respond to mainstreaming policies from situated locations and have internalized historically produced understandings of the "common." In this regard, I invite the reader to assess the role of the state in the contemporary manipulation of these maternal narratives through the use of two analytical themes: "disciplining" and "panopticism."

The process of "disciplining" functions as the force through which the social order institutes power relationships. We are disciplined into a number of socially sanctioned locations through a constructed belief in and adherence to narratives of normality, abnormality, and dysfunction. If acts of discipline are achieved through a range of competing social narratives about propriety, morality, and normalcy, then what are the discursive narratives that bring Caribbean women into relationships of compliance, complicity, and desire for the sanctioned narratives of womanhood as enunciated within state machinery? This line of questioning aims to identify and interrogate the state practices (both colonial and nationalist manifestations) that propagate what Jacqui Alexander refers to as a variety of "fictions of feminine identity" that work actively to align female desire through terms that are compatible with national narratives of femaleness (1994, 140).

The political desire that informs these "fictions of femininity" is the creation of a docile body—to produce a docile maternal body that lacks the capacity to react or challenge and one that makes it difficult to enunciate different gendered possibilities for women within mainstreaming approaches. Speaking of the docile body in relation to the effect that "discipline" produces, Foucault observes:

> Discipline increases the forces of the body (in economic terms of utility) and diminishes these same forces (in political terms of obedience). In short, it dissociates power from the body; on the one hand, it turns it into an "aptitude," a "capacity," which it seeks to increase; on the other hand, it reverses the course of the energy. (1995, 138)

Enervation, therefore, is the ultimate goal of docility. In other words, the successful production of a docile maternal body is through the maximization of its economic returns (e.g., as an electoral resource when seeking votes) and the minimization of its revolutionary capacity (e.g., dependency on state resources). While I focus on these processes in the context of social services because of its institutional connection with gender mainstreaming units, we need to note that these processes of subject making are not linked to any one institution or apparatus in society. Rather, these narratives are embedded within a range of specialized institutions, such as schools, hospitals, prisons, and welfare agencies. These institutions are invested with the force to ensure adherence by those who come into contact with them via an internalization of the institutional narratives that exist about the various bodies who use their services (e.g., working class students and widows).

Within Foucauldian thought, the panoptic functions as the ultimate performance of surveillance. It is the omnipresent Eye/I that encourages us to align our behavior and identities with normative positions or subject locations for fear of being caught outside of the designated codes. Foucault's use of the "panopticon" is based on Jeremy Bentham's architectural surveillance innovation for maintaining control over prison inmates. Foucault notes that the major effects of the Panopticon are:

> ...to induce in the inmate a state of conscious and permanent visibility that assures the automatic functioning of power. So to arrange things that the surveillance is permanent in its effects, even if it is discontinuous in its action; that the perfection of power should tend to render its actual exercise unnecessary; that this architectural apparatus should be a machine for creating and sustaining a power relation independent of the person who exercises it; in short that the inmates should be caught up in a power situation in which they are themselves the bearers. (1995, 201)

Architecturally, the Panopticon was a centralized fortress in the middle of surrounding prison cells. The surveillance power of the Panopticon lay in its unverifiability—there was no way of knowing whether there was a guard on duty in the tower or not. Consequently, the Panopticon's effectiveness resulted from its ability to produce self-surveillance, a sense of constantly being watched to the extent that the subject internalizes this monitoring capacity within herself.

The effectiveness of the Panopticon is located in the fact that it is not a person or an institution, but rather, a functioning of power. As such, I examine how the exercise of state power can come to function in this panoptic way in that any individual, a social worker, a next door neighbor, or a child, can serve to provide state practices with its immanent force by inducing a sanctioning capacity over women. Rather than a disinterested formation, we can then unmask the agendas and stratagems that aim to produce and secure for its own purposes a docile, maternal body (or any other gendered narrative). I also wish to examine how the women themselves internalize and become resistant or complicit with these scripts.

The formation of the Trinidad and Tobago national machinery, the Department of Social Services, is important both as the Ministry with the longest institutional history (July 1939) and as one of the primary avenues through which women have been incorporated within state machinery and the one that has been most closely aligned with gender mainstreaming divisions, where many maternal specters come to reside. This Ministry (differently named over time) provides critical insight into the state impetus toward the disciplinization of the female body within state machinery. It is important that we read the performance of this ministry and all state formations for their panoptic character, as an instituting force on the formation of a socially sanctioned maternal subjectivity.

Public Assistance Act, Chap. 32:03, Act 18 (1951)[21] governs the administration of Public Assistance. As with all legislation, the recipient of public assistance is referred to "generically" as male. More importantly, there is no clear definition of the term "Head of Household,[22]" and, further, there are no cultural nuances that reflect the supposed centrality of women in Caribbean family forms. In practice, however, that which is written generically is enacted with surgeon-like gender specificity, and in practice a woman will not be considered or designated as head of household if there is an adult male partner domiciled within it.[23]

This definitional anachronism has both private and public implications, that allow us to see welfare clearly as an inherent attempt at managing the female body, both by way of erasure when acting on behalf of others or as a dysfunctional financial excess when acting on her own behalf. The narratives legitimize the erasure and devaluation of the "headship work" (e.g., income generation, decision making, disciplining, household maintenance, and management) that women do, indeed, perform within the domestic sphere even when a male may be present. Yet, the lack of political will to employ a more culturally relevant and gender sensitive definition of the category "head of household" reinforces the state's refusal to deal with women as agents in their own right. Instead, this refusal privileges the *a priori* right of the male to deal with the state, as a dialogue among equals. This gesture marks an impulse towards further discrediting the kinds of negotiations that women conduct when acting on behalf of children and/or incapacitated male relatives and partners on welfare. Women remain subordinate to both the institutional culture of state machinery *as well as* the gender-based power relations of the domestic sphere as a result of the state's conceptual and operational enforcement of a male head.

Unyielding adherence to the "male head of household" script ensures that the organizational force of the welfare approach re-imposes the colonial desire to secure a nuclear originary as the basis of the state. The application of the term "head" simultaneously invokes a number of other designations, such as "worker," "income earner," and "provider." However, this also brings to the fore the problematic corollary of "not nurturer," "not emotional," and "not domestic." These conceptual limitations and tropes align the state with nuclearity and enforce traditional gender roles within the domestic sphere. Women who may be the primary income earner are disqualified from receiving any state assistance if they can no longer earn an income as long as there is an able-bodied male within the household. There is, consequently, a devaluation of women's economic activity, as well as the emotional value given by both partners.

A series of events that open women to a number of state-sanctioned intrusions are set in motion when women dare to approach the state and declare themselves as household head with the primary or sole responsibility for children. Having presented themselves as agents before the state, a number of practices are initiated with the overall intent of making the maternal body docile. A number of actors, such as neighbors, children, and social workers, are all co-opted into the surveillance and disciplining of what can only be these women's already wanton sexuality. This sexual surveillance reminds us of Alexander's observation that the state invests in resources, machinery, and will toward the denial, suppression, and erasure of women's sexual agency. The fiction that must be perpetuated is that of the nuclear originary and women's bodies, and their self-conceptualization will be made malleable to this image (1997, 64).

It is not surprising, then, that the state-induced processes of disciplining the maternal body incorporate control over maternal sexuality, seen most readily in the procedures that women must follow to receive assistance from the state. While a male adult simply has to declare himself as medically unfit to secure services on behalf of himself or children, women must declare and account for their sexual past before the state, which becomes the patriarchal head by proxy. State-agents, such as welfare officers, functionaries of the court, and the medical professional who declares a woman medically unfit, must all first ensure that they are not supporting a deviant form of masculinity or that they are allowing the state to be "cuckolded," otherwise known as "unnecessary public expenditure." They accomplish this verification in an apparently innocuous manner, according to the following guidelines given by a state-based social worker:

> The single adult is medical, the children if it's a single woman because we have men who apply for children too, eh, but the men have to show that they are certified medically unfit, but the single woman applying for children they have to submit the birth certificate. They need an affidavit saying who is the father of the child or fathers of the children and, um, when they've proved that the father is who the father is, then we have to find out what happened to the father. If the father is dead, you produce the death certificate. If the father has deserted the home, we have to find out if the person went away. If the father went away, they have to prove that the person is out of the country by submitting a date that the man left Trinidad. We will send the date to immigration, asking them to verify that that is the date that the person left. That is a long process, but eventually, that is our proof that the person is out of the country, so we have to wait on that before we could process the application. (interview with Social Worker I May 9, 2002)[24]

Yet, when state agents, such as welfare officers, ask women to identify their reproductive history, they simultaneously map a moral geography that requires women to account not only for the existence of their children, but also the whereabouts of the father(s) of these children. This moral geography, once mapped, opens up women's domestic sphere to the panoptic gaze of the state. Consequently, even though the woman has presented herself and/or her children as necessitous, she must first account for the state's anxieties about paternity and, by extension, her own moral culpability in the present dilemma.

Acts of surveillance are, therefore, deliberately designed to curb maternal sexuality in ways that suggest welfare to be an act of state-induced neutering. Following on the processes that require women on welfare to chart their moral geography is the activation of the panoptic through unexpected and immanent acts of surveillance that ensure that all agents are potential surveyors with disciplinary capacity: the unwitting child who may be asked if mummy had a man over last night; the social worker who may ask such a question; or the neighbor who might deliberately or unwittingly respond. This stands in direct opposition to the state agent's representation of this relationship. Two different but very interesting narratives emerge. The first is the need to protect the state from unscrupulous requests. Related to this is the fear of supporting non-performing or deviant masculinity. In this way, welfare policies bring women into a complicit relationship with the state in so far as they are now invested with the responsibility of disciplining deviant masculinity if their own needs are to be met. The respondent continues:

> . . . We have to prove that the woman has tried to get maintenance from the man because it was a mutual thing they had this child. So you have to prove that they're trying to get some assistance from the man before they come in. But you find that the man would be in the home and they would still come and they would say the man isn't there and we can't be there *twenty-four* hours since they will know when we're coming. We don't know the man, so we are in a position where they could say that is a friend and we have to take it because *we taking people word*. (emphases mine; interview with Social Worker II May 9, 2002)

Interestingly, the very occurrence of the visit indicates that the system is not based on "taking people word." On the contrary, documented proof—for example, police certificates if imprisoned, migration documents if abandoned, and death certificates if widowed—must be provided by the client at each stage of the transaction.

Maternal subjectivity is treated as suspect, requiring documented evidence to prove otherwise. Item 8(a) of *Application for Public Assistance* requires documented proof to corroborate all claims. Clients must present all necessary documents; these must be signed by the state agent as having seen them and comments made regarding the legitimacy of the claims. The application form also requires that site visits be made by the assigned agent, again to ensure the legitimacy of the client's claims. At each stage, checks are instituted to ensure that the system is not required to take the client's "word." This overdetermined assimilation of the state's unarticulated ideological intent can be heard, for example, in a social worker's declaration that the state cannot "spoil the man," and, consequently, "that is why we kinda pressure the women, make it harder for the women to come in because they must have proof."

The use of the term "spoiling the man" suggests a woman's indulgence of male non-performance in traditionally masculinized roles. However, the case above makes very clear that the state endorses and sanctions a very clearly demarcated gender continuum. As such, as women's productive work in the domestic sphere is discounted, men's emotional work is similarly discounted. Women, though, become the scapegoats in this arrangement (*That's why we pressure the women, make it harder for them to come*). Imposing a greater degree of opposition for women to approach the state vests the woman with the responsibility for disciplining "her man" into appropriate forms of masculinity or holds her liable for not having done so.

In those cases where mothers take fathers to court for maintenance as a procedure of producing "proof," there is the dimension of public shaming with women's sexuality being placed on trial. In the absence of sophisticated DNA paternity tests, decisions regarding paternity invoke all historically derived scripts of maternal morality in which a woman's credibility and claims for paternal assistance can be compromised by assertions of sexual promiscuity. Not surprisingly, women are hesitant to pursue this process because of the extent to which they, rather than the absconding father, are placed on trial. Women's ready awareness of the potential to undergo acts of public "shaming" when initiating legal procedures to claim rights and entitlement in the area of domestic violence or child support resonates with the conclusions of Mindie Lazarus-Black's work, "The Rites of Domination: Tales from Domestic Violence Court" (2002).[25] Based on her ethnographic research in the Magistrates Courts of Trinidad and Tobago, Lazarus-Black observes that many of the legal processes that women undergo when attempting to claim protection or support further disempower rather than empower those who are subordinate. Processes of shaming, or what Lazarus-Black refers to as "a rite of humiliation," captures the stratagems that are used to attach stigma to women when they challenge the state or other forms of patriarchal power. The women who participated in this study understand these processes of "proof production" and, when possible, guard against having to present themselves before the courts where they would be compelled to publicly expose intimate details.[26].

State skepticism about the veracity of women's testimony requires constant disciplining and surveillance to ensure that the maternal body is a traceable one. Inherent in the need to secure "proof" is the desire to achieve "twenty-four hour" surveillance, and one strategy of achieving this is through unannounced home visits—the primary way in which this surveillance is performed. The mother is constantly aware that, by virtue of being on public assistance, there is a measure of domestic permeability to state agents and, despite the fact they sometimes announce when they might visit, they may easily as well not do so.[27] Take as an example the following interaction that occurred when a social worker visited one of her client's homes early one morning:

> We had a supervisor who used to actually send you early in the morning, to look for men's clothing and men's shoes. So at one time in one of those plannings, I went to a woman. I remember her name up to this day. She had a lot of children and I'm talking to her and I'm seeing something on the table shifting and I say, "But 'C,' yuh have rats, yuh have to be careful. We have to get the pest control people . . ." She say, "No, what yuh talking about is just the breeze." I say, "C, no, no, no, no . . . Uncover this thing and let's see what under there." And when I uncovered it, it was a child. A new born baby. When she heard me coming, she throw the plastic oil cloth, plastic eh, over it to hide it because she didn't want me to know that she make a next child because when they make another child they used to stop they money at that point in time because it means the father there. (interview with former Social Worker III/feminist activist May 9, 2002)

The main disciplining agent is the state's capacity to withdraw financial assistance for any perceived infringement. In this narrative, the mother's internalization of this disciplinary component of the state led to instances of self-surveillance and acts of subterfuge that bordered on the surreal. Despite the fact that this mother had interacted with a social worker who, over time, had exhibited that she understood that both financial assistance and sexuality were component parts of her maternal subjectivity, the recipient had internalized the regulations to the extent that she saw it necessary to endanger her baby's life to keep assistance.

This manipulation and control of women's bodies and sexuality prevails throughout the institutional framework of the Trinidadian state. This was particularly evident with the following respondent. Joan, a single mother of five, was abandoned by her spouse fourteen years ago. News of her final pregnancy took her completely by surprise because both she and her husband had authorized the hospital to perform a tubal ligation after the birth of her fourth child. On learning of her pregnancy, Joan remembers:

> Girl, I went up in Family Planning and I did a pregnancy test and it came back positive. Well, my blood pressure just went up and I was like, I fly across at the hospital one time. Gone across on maternity, all the nurses had to calm me down, you know. I don't curse and get on an' ting. And you know they will pick up for the doctor . . . Well, the nurses start calling for the doctor and they say, "Well, that doctor he doh (don't) like to do that ting," and start covering. I said, "But the papers were signed," and when they checked back meh file, they see the

papers were signed. I sign, meh husband signed. It done happen and she was the prettiest baby I ever had. (interview with welfare recipient May 11, 2002)[28]

Joan's narrative is filled with a sense of impotence and a sense of disenfranchisement that women often experience when interacting with state-empowered agents. Strikingly, Joan did not express regret because of the fulfillment that she links to childbearing *(It done happen and she was the prettiest baby I ever had)*. However, and in no contradiction to this narrative of self-sacrifice, did Joan articulate any sense of entitlement from the state for a clear breech of ethics and malpractice on the part of the doctor. This narrative normalizes compatibility between maternal self-sacrifice and disenfranchisement. The doctor as a state agent, as a matter of personal beliefs, produced a situation in which Joan's reproductive rights became predicated on the veto power of the state. With neither the confidence to assert her rights at the indignity, nor the financial wherewithal to pursue the matter legally, Joan, at that time unemployed, had the baby. Her husband abandoned her two years later.

When asked if she pursued assistance from the state after her husband's departure, she responded:

> Somebody tell meh, "Joan, why don't you go to Welfare and apply for help?" They say that is money. You work and put money there, you entitled to that and eventually I really took the advice. But the kind of things they have to know, the amount of questions they have to ask you before you could get that money. And they really came home, they had to see where I living. If you had T.V., if you had radio, they want to see everything you have because they find that if you going on Welfare, you shouldn't have these things. This kinda way. And a girlfriend was telling me. She say, "Joan, when yuh go into them, you can't go dress up and looking nice." Ah say, "I am not going looking anyhow." She say, "Yuh cyar do up yuh hair because if yuh do up yuh hair . . . . they go find . . ." "Well," I say, "too bad for them, I not begging them for anything." That is my way. (ibid.)

The process of securing welfare or assistance systematically sets about to erase any sense of agency and empowerment in the maternal body. While this respondent challenges the right of the state to impute value to her material and aesthetic well-being, other women interviewed attempted to respond to this by muting any measure of aesthetic appeal. In other words, the system produces the need to perform poverty in order to receive assistance. In this performance of poverty, appliances that may have been acquired over time, markers of beauty, self-esteem, confidence, and attractiveness are all part of forbidden narratives attached to this maternal reality.

Docile corporeality becomes the subtext of state-imposed discipline. A number of gestures work toward producing this docile corporeality. From their very first visit, mothers are asked very intimate questions about their sexual history and are compelled to answer. The inappropriate becomes acceptable, necessary, and acquiesced to, as a rite of passage to financial assistance. The architectural design of the meeting point denies privacy, as one can be heard from cubicle to cubicle, and very intimate details of one's life become the possession and purview of the public domain. As we will see, women find a number of ways of subverting the system. However, it is evident that structures and procedures are designed to achieve increasing gestures, or at least *performances* of docility, the longer women stay on welfare.

This docility was strikingly evident when interviewing women on welfare within the agency. The ease with which interviews were secured was in part due to the inherent authority of the research process itself with its tape recorder, a student with the officious "foreign" identification card and letterhead, and a missionary desire "to help." However, the environment itself carried its own history and authority of interrogation. The process of waiting, the awareness that your exchange can be heard since you yourself can hear the exchange in the next cubicle, and the forced exposure that is only partially counteracted by responses voiced in hushed tones or seemingly sheepish smiles that attempt to indicate politely, if not deferentially, that some questions were not going to be answered, and the fact that I deliberately respected these smiles and did not probe or pry if a question was not answered when first asked—all of these features expose the intimate and leave little by way of privacy. Yet, the intriguing aspect of interviewing in this locale is that it allows us to assess the ways in which clients assume the persona of the "woman on welfare." Yet, as the earlier narrative that recounts the mother's attempt to conceal her baby from her caseworker would suggest, there are always attempts to resist and challenge these institutional attempts at control.

What kinds of statements have we come to expect about Afro-Caribbean women's identities? Who and what possibilities are foreclosed by these enunciations? How might these enunciations and foreclosures affect what is possible in the project of gender equity? I offer this chapter as a methodological and conceptual response to these questions. The epigraph that begins this chapter admonishes us to attend to the historical processes of apparent unities so that we might better understand the inner workings of potentially repressive subject representations (Scott 1991). It may be naïve of me to expect all equity projects to proceed with the same kind of genealogical analysis that I have offered here. My larger conceptual point, though, is a legitimate one: gender relations pre-date the arrival of gender manuals and workshops through which prescriptive notions of gender are disseminated. So that we understand how local conceptualizations of "gender" as a category of analysis might be received, obstruct or be problematically compliant with the "gender" of mainstreaming projects, we must at the very least invigorate both a disciplinary and methodological conversation that allows us to understand local processes of subject formation and how these have been institutionalized in ways that name some as worthy within state-sanctioned constraints and others as unworthy.

In this chapter, I have experimented with an analysis of how historicized processes of Caribbean gender relations congeal and conceal what it is possible to say about women's subjectivities. The narratives that have been consolidated around Afro-Trinidadian women's bodies through processes of (M)Otherism, institutionalization of social crises, and disciplinization of women's reproductive capacity hover as already embedded within the mainstreaming, equity, and advocacy issues that are now most pertinent to the region, namely, reproductive equity and sexual minority rights. Feminist advocacy, as I will next argue, must create new forms of intimacy with these foreclosed, at times abject, spaces if it is to credibly take the lead in envisioning a new gender order within the Anglophone Caribbean.

3 Co-Opting Gender and Bureaucratizing Feminism
Exploring Equity through the Institutionalization of "Gender"

The proliferation of international gender-based documents, a plethora of gender indicators, and the eagerness of states—particularly in the Third World—to ratify these documents as an avenue toward increasing their credibility on the global playing field makes it difficult to sideline issues of women's empowerment, rights, and gender equality from the discursive landscapes of national policy discussions. Yet, how does one account for the fact that three decades after the First World Conference on Women in 1975, Caribbean women, in reality, appear to be under the same and, in many instances, even greater threat within the planning machinery of these territories? In this chapter, I address these questions by analyzing the dissonance that exists between the abundance of institutional mechanisms within the region and the experiences that gender focal points have on the inside of these institutions.

My curiosity here is to examine the contradictions between the appearance of a regional institutional framework mandated to realize the goals of gender equity and the internal administrative hurdles that impede the realization of gender equity, both often emanating from the same spaces and persons. Throughout this chapter, I also draw attention to the psychic toll that implementing an equity agenda has on those designated as "femocrats" as they work within the interstices of these contradictions.

My discussion here is a scaled one. I begin with an overview of some of the common weaknesses that now inform our understanding of implementing mainstreaming policies globally. I then turn my attention to the regional level to assess the institutional strengths that exist in the Caribbean. I follow this with a case study in which I analyze interviews with state functionaries from the Trinidadian Gender Affairs Division, the lead gender focal point within the nation's machinery; these interviews capture the institutional memory of members of the ministry during critical periods within the Division's development. By shifting analytical scales, we are better able to highlight the rhetorical sleight of hand that now allows state-managers to declare that women are centrally located within planning and policy formulation because institutional frameworks exist. This is an incongruity that points to the frame while masking the absence of content and has to be unveiled as a *form of heteropatriarchal resistance* within frameworks of existing institutions responsible for equality mandates.[1]

MAINSTREAMING EQUALITY GLOBALLY: INSTITUTIONAL CONSTRAINTS, PITFALLS, AND POSSIBILITIES

With the exception of the region's history of feminist legal advocacy, gender mainstreaming stands as the primary vehicle through which discussions of gender equity occur in the Caribbean region.[2] Globally, there are few exogenous protections for mainstreaming processes (e.g., law, policy, and procedural operations) in most countries. This is particularly so in the Anglophone Caribbean. Gender mainstreaming, despite being lauded as a vehicle of women's advancement, holds neither pride of place, nor is its mandate protected within the state. Consequently, in the absence of accountability mechanisms, there are common institutional vulnerabilities that have begun to emerge globally.

Gender mainstreaming's institutional precariousness threatens to undermine any sustained commitments to change. As a form of state feminism, gender mainstreaming's success is critically linked to its institutional location and the authority that resides in the office or officials with whom implementation rests (Goetz 2007).[3] The model is extremely reliant on the patronage of individual agents and the twists and turns of ministerial reallocation. It is vulnerable to the vagaries of political expediency and individual leadership idiosyncrasies, the unpredictably of which undermines its stated vision of gender equality.

Former Director of the Women's National Commission, Janet Veitch's "Looking at Gender Mainstreaming in the UK Government" offers an important overview of the U.K.'s mainstreaming context from 1997 to 2005 (2005). Veitch's discussion shows a mainstreaming agenda besieged by cabinet shuffles, by subsequent changes of ministers, ministry, and policy emphases, and by a lack of appropriate expertise among the individuals with responsibility for implementing a mainstreaming agenda. Drawing on Teresa Rees' idea of "tinkering" and "tailoring," Veitch notes that the unevenness of implementation processes has resulted in some policies that "tinker" with existing policies and laws to bring women in line with opportunities for men, and in some strategies "tailored" to address women's specific disadvantage as a group[4] (Rees 2005; Veitch 2005). The United Kingdom's lack of consistent high level support, Veitch observes, produces gaps in terms of resources, strengths, and expertise that limit the transformative capabilities of mainstreaming.

"Gender's" conceptual ambiguity also results in very different political valences within the global context. Australian scholars Carol Bacchi and Joan Eveline's comparative discussion of Canada and the Netherlands shows that different understandings of "gender" led to significantly different operationalizations and political commitments at the state level. The Canadian context placed an emphasis on achieving gender equality through women's difference, an approach that Bacchi and Eveline conclude led to providing women with resources that alleviated their disadvantage without necessarily confronting the assumptions of the gendered status quo (2005, 505). On the other hand, the Netherlands, they argued, deployed an understanding of gender that centered the asymmetry of gender power relations. Bacchi and Eveline observe that by de-centering the "male norm," the Netherlands has been able to bypass the "sex/gender distinction [to] an understanding of gender as a political process" (2005, 507).[5]

An assessment of selected mainstreaming contexts within the North/South divide yield unexpected results. For example, feminist economist Carolyn Moser, after assessing the effectiveness of Northern donor agency work in the global South, concludes that the South may in some ways have proceeded farther along than the North thought the South had in challenging gender neutrality at the state level (2005, 588).

Practitioners in the South paint a more testy picture. Law professor L. Amede Obiora's piece "'*Supri, supri, supri, Oyibo?*': An Interrogation of Gender Mainstreaming Deficits" (2003) begins with a vignette in which a little girl encounters a white foreigner in a market in Kampala. Having designated herself as a translator for the foreigner, she proceeds to communicate using a kind of gibberish that she imagines to be the foreigner's language. Obiora argues that this incommensurability between the "foreigner's" language and the young child's gibberish characterizes the mainstreaming frameworks that travel to the global South. Within the South, global economic vulnerabilities are played out as a struggle of expediency between discourses of resources and rights. Whether one uses scarce resources to hold a workshop to improve women's economic livelihoods or to initiate legal changes is lodged at this interface of resources and rights. Obiora also points to the ways tropes about "African women" can lead to paternalistic policy formulation within international organizations, where choices about resources and rights are guided by these very tropes.

With governments ever contracting and weakened by terms of the global political economy, the NGO infrastructure in the global South is of increasing importance to women's advancement. This framework is not without its own implementing constraints. Researchers Senorina Wendoh and Tina Wallace in "Re-thinking Gender Mainstreaming in African NGOs and Communities" draw on their research done with women's NGOs in Zambia, Rwanda, the Gambia, and Uganda. They point to an NGO infrastructure overwhelmed by a constellation of connected vulnerabilities (HIV/AIDS, poverty), rural/urban hierarchies, the perception of mainstreaming as a foreign imposition, and conceptual uncertainty regarding "gender equality." Additionally, social inequities of class, ethnicity, and tribe within the broader society, when replicated within the NGO community, impede mainstreaming efforts. Individually, each of these factors would be serious enough to weaken a mainstreaming approach. Together, they produce confusion and hostility (Wendoh and Wallace 2005).

The NGOs' dependence on government subsidies and international donor agency grants also imposes a degree of artificiality on the issues of importance, the timeline for implementation, and the indicators of success that NGOs are expected to comply with as a funding requirement. Wendoh and Wallace note that in these contexts the "challenge for donors, governments, and NGOs is to find ways to support and encourage positive change in favour of women, rather than bringing in blueprint ideas and concepts that have no meaning for local actors" (2005, 79). This process of indigenizing and translating mainstreaming's relevance to the local context continues to be one of the most important and difficult aspects of securing a sustainable equity agenda in the global South.

Unlike most accounts of mainstreaming, Hester Eisenstein's ethnographic study of Australian femocrats in *Inside Agitators: Australian Femocrats and the State* (1996) paints an optimistic picture of the accomplishments that are possible when those with a feminist sensibility are able to confidently maneuver within state machinery. Eisenstein identifies the 1980s as the period in which more positive meanings came to be associated with the role "femocrats" played within government. Although earlier references to the term would evoke a cadre of feminists who had either been co-opted by the state or succumbed to the lure of capitalism by securing jobs within state machinery, by the 1980s, particularly in the Australian context, the term more readily painted an image of a "powerful woman in government administration, with an ideological and political commitment to feminism" (68).

The contrast between Eisenstein's Australian study and my own discussion of the Caribbean drives home the importance of understanding gender mainstreaming processes as highly particular and contextual phenomena. Despite the prevalence of UN directives, manuals, and workshops suggesting best practices, the successful implementation of mainstreaming agendas is dangerously vulnerable to contextual caprices. Eisenstein identifies a number of successes within the Australian national machinery that she credits to the influence of "femocrats" who forged a network for high-ranking feminists among policy makers, using their influence and resource access to strengthen the wider women's movement, centralizing women's issues for consideration at all levels of policy and fiscal decision making, and changing attitudes and behaviors within the wider society (43–44). The femocrats' partisan affiliation with the incumbent political parties also contributed to the influence that certain "femocrats" held. This party allegiance opened a degree of state responsiveness to feminist efforts and the eventual recognition of gender mainstreaming as a form of expertise in its own right. However, Eisenstein notes that none of these successes were a given, and she further identifies a number of hindrances experienced by the femocrats in her study. These included resistant bureaucratic culture, constraints on the femocrats' authority, undue political partisan influence, and opposition from those who benefit from the existing gender status quo (170–177).

Similarly, Jo Beall's comparative discussion of mainstreaming attempts in Colombia and South Africa would suggest that while partisan political support of incumbent parties is critical, the potential of implementation can quickly plummet if the individual patron loses credibility or favor with the party or wider civil society. This, Beall suggests, certainly affected the success of gender reforms in Colombia under the 1994 regime of Ernesto Samper Pizano. In contrast, in post-apartheid South Africa, gender equity was deemed important enough to merit constitutional protection. Still, constitutional arrangements are not a panacea, and Beall continues to emphasize the importance of consensus and coalition building in the pursuit of challenging long-held traditional beliefs about the "rightness" of male supremacy (1998, 528).

Yet, strangely missing from the many international guidelines for effective mainstreaming efforts is the importance of a non-partisan commitment to the philosophical premises that underpin the technical processes of mainstreaming. Such a non-partisan agreement would incorporate gender equity as part of the ethos of governance, consequently avoiding the five-year wind of political/electioneering expediency. This would warrant profoundly important conversations about what justice, equity, and dignity mean in national contexts, and it may well be worth the time and resources of international donor agencies and change agents at the national level to serve as catalysts for such a conversation alongside the current emphasis on technical expertise and procedures that now dominate present mainstreaming discussions.

The multiple sites of dissemination, as well as the reinforcements that are found in subsequent international documents, contribute to persistence and prevalence of mainstreaming approaches internationally. Throughout the 1990s, in anticipation of, and later in response to, the Beijing World Conference on Women, a number of feminist practitioners invested significant energies into setting the conceptual and institutional best practices for gender mainstreaming. These were later reinforced by the UN's momentous release of gender-based indicators—Gender-related Development Index (GDI) and Gender Empowerment Measure (GEM)—in the 1995 *Human Development Report*.

With a degree of optimism, not misguided because of the nearly universal adoption of the Beijing Platform for Action, feminists worldwide saw gender mainstreaming as the most viable way of incorporating gender equity concerns into state operations. Almost fifteen years later, gender mainstreaming has undoubtedly had limited success in achieving long-term gender equity. Feminist development practitioners from different disciplinary vantage points worked hard at identifying appropriate strategies, concepts, and mechanisms that would facilitate gender-just change through mainstreaming processes (Staudt 1990; Rathgeber 1990; Moser 1993). The multiple sites of dissemination, as well as the reinforcements that are found in subsequent international documents, contribute to persistence and prevalence of mainstreaming approaches internationally.

Political scientist Rounaq Jahan's *The Elusive Agenda: Mainstreaming Women in Development* (1995) was one of the earliest attempts during this period to provide readers with comprehensive schema of the strategies needed for effective gender mainstreaming.[6] Her institutional strategies itemized the importance of allocating responsibility throughout the institutional framework. She crafted a multi-tier framework that required an exchange of accountability between different components of civil society and government, and she supplemented monitoring and evaluation procedures with the need for establishing a personnel policy that would strengthen the level of expertise required and diversify the labor market where appropriate (37). Similarly, Jahan's list of operational strategies emphasized the need for clear guidelines at every implementing level, enhancing personnel's capacity, attention to generating appropriate statistical data and research, along with country-wide programming and sustained gender analysis. Taken in tandem, Jahan's institutional and operational schema provided systematic strategies, which, if implemented, would challenge daily institutional operations and procedures to incorporate more gender-sensitive approaches.

Although studies like Jahan's seemed initially to identify a definitive model that could be used for institutionalizing gender mainstreaming in other countries, feminist practitioners found them inadequate to the enormity of the challenges they faced. The lack of significant inroads to gender inequity globally prompted feminist practitioners to question the analytical emphasis on institutional analyses, which at times too readily privileged process over outcomes (Staudt 2007). There was agreement that evaluation and monitoring mechanisms needed to be more robust if there was to be any hope of stemming the tide of what Moser has referred to as "policy evaporation," that is to say, the gradual disappearance of gender-related policy from the agenda (Moser 2005; Staudt 2007).

Nevertheless, there continues to be an emphasis on methodology within gender mainstreaming, sometimes to the extent that process may at times run roughshod over person. I am in complete agreement with political scientist Kathleen Staudt's observation that to fine-tune gender mainstreaming's implementation and outcome possibilities practitioners

> must understand how people and institutions are *gendered*; that is, they must address, compare and assess their missions and programs for burdens and benefits on women and men in the given asset and opportunity structures of unequal power relations. (2007, 46)

I find Staudt's observations useful on two counts. First, understanding how people are gendered enables us to re-center the question of gender asymmetries, an issue that can sometimes get lost in our preoccupation with mainstreaming methodologies. This need to understand how people are gendered also allows us to recognize the multiple layers of gendering that inform a more complex understanding of gender mainstreaming, in contrast with the now dominant binary modes of heteronormative masculinity and femininity.

Second, Staudt's observation that institutions themselves are also gendered is one that I will engage with some detail in this chapter. It is not helpful to think of institutional gendering as merely "masculinization," as reflected by numerical dominance of male bodies or by aggressive, boorish characteristics. These have certainly been legitimate characteristics in the literatures of gendered labor markets or the performances of hegemonic masculinities (Elson 1998; Connell 2005). Institutional gendering manifests through a potentially coercive stance in the maintenance of the gendered status quo. Still, this is not a project with which only men are complicit. If we see the gendering of institutions in only this way, then we have simplified a number of realities that consolidate to marginalize a number of different constituencies within Caribbean societies.

As I argued in Chapter 2, to fully understand the gendering of institutions, we should undertake a practice of historical analysis, particularly for those institutions given contemporary oversight for gender mainstreaming. Historical analyses also help us understand the ways in which various categories of women and men have been codified within the state. Still, the gendering of institutions should also address the ways in which a bureaucracy demands the performance of certain gender identities from those who solicit its protection, support, or good will. Commanding gender performances such as the "good mother," the "worthy victim," or the "delinquent father," whether through policy or informal culture, should moreover be seen as an integral part of the gendering practices of state institutions.

It is difficult to discuss the gendering of institutions without acknowledging the profound influence that sociologist Max Weber's rational-legal ideal type bureaucracy has had on shaping our understanding of how bureaucracies "should" work. A proviso is necessary here: I am by no means re-invoking a male/rational, female/emotional dichotomy. On the other hand, if, according to Weber, a bureaucracy develops "... more perfectly, the more it is 'dehumanized,' the more completely it succeeds in eliminating from official business love, hatred, and all purely personal, irrational and emotional elements which escape calculation" (Weber 1978, 975, cited in Albrow 1997), then what do we make of gender mainstreaming procedures that attempt to centralize human worth and incorporate less hierarchical management styles? Can these be reconciled within a bureaucratic ethos?

The dichotomy here is not that of male/irrational or female/emotional. It is, however, a more complex analysis of the constraints that feminist practitioners face when attempting to implement models or ways of being that do not adhere to the dominant mode of operation. In other words, when various constituencies of women and men contravene this dominant mode, they will be sanctioned, and the nature of the sanction will come through a deployment of the terms by which they have already been gendered (e.g., men become "soft" and women "poor decision makers"). This is somewhat different from positioning an always-already rational male/emotional female in favor of analyzing institutional cultures with an eye for the gender identities that both women and men are called upon to inhabit.

With these observations in mind, I want to turn our attention from the global to the regional in order to examine the institutional framework that exists for gender mainstreaming in the Caribbean.

SETTING THE AGENDA FOR GENDER: A REGIONAL OVERVIEW

Anne Marie Goetz (2007), in a five-country overview of gender mainstreaming, examines the effectiveness of national machinery in the respective territories.[7] Among the five territories, Goetz identifies three institutional models of mainstreaming. These include advocacy units that are located in central planning units and are invested with lobbying responsibility. Advocacy units, while not institutionally independent, are vested with responsibility for lobbying, increasing awareness of and sensitivity to gender issues. Second in her typology are units with oversight responsibility. These are monitoring units in which the process of gender mainstreaming occurs through monitoring and evaluative assessments of projects and policies emanating from central planning agencies. Finally, her third model is implementing units; implementing units serve an exemplary function for other ministries within state machinery. They have programmatic capacity and take lead responsibility for gender issues not addressed elsewhere within the "malestream" political framework (e.g., domestic violence and gender sensitization; 71–72).

An overview of the region's institutional framework shows that advocacy and implementing units are the predominant gender mainstreaming models. Regardless of the institutional type and organizing rationale, as a mode of equity gender mainstreaming is precariously positioned within the overall governance framework of the region. One of the earliest assessments of the region's institutional strength on women's advancement paints a very dismal picture. Peggy Antrobus, founding member of DAWN and Advisor on Women's Affairs to the government of Jamaica in 1974, noted in 1988 at the Inaugural Seminar of the University of the West Indies (UWI) Women and Development Studies Project that the national women's machineries "remained under-staffed, w[ere] transferred from one ministry to another, or ignored, and largely ineffective" (1988, 43). Quoting from the Commonwealth Secretariat study of National Machinery in the Caribbean, Antrobus went on to acknowledge that the region's bureaus and units had successfully raised the region's awareness on issues related to women's welfare but that there remained a significant gap between addressing these concerns and linking these concerns to their structural causes (ibid.).

Sixteen years after Antrobus's daunting assessment, the region's gender mainstreaming practices continue to manifest a number of similar constraints and hindrances. In interviews conducted for the Fourth Ministerial Conference for Women in the Caribbean (2004), gender focal points in the region identified the following as some of the more dire constraints to effective gender mainstreaming processes[8]:

- An absence of sex-disaggregated statistics to inform research, programmatic planning, and policy;
- Little or no conceptual understanding of "gender" and gender-related issues among mainstream technocrats;
- Dire need for gender sensitivity training for staff and other technocrats;
- Increased allocation of resources for implementation purposes and project management;
- Marginal relationship to other ministries in the review and critique of policies that are to be implemented;
- That "male marginalization" is increasingly named as the urgent gender issue; and
- Little political commitment given by senior officials and Ministers.
  (UNIFEM/Rowley 2003, 8)[9]

"Gender focal points" refer to individuals with designated responsibility for technical implementation, advisory, monitoring/evaluation, and training and/or support capacity for mainstreaming processes. The responsibilities placed on these individuals vary with the institutional context in which they work. As the only person within an advocacy unit, they may be the central point of contact for all things gender related; they may be housed in an implementation unit, one staff member among others; or they may follow a more decentralized format and reside in their respective ministry with monitoring responsibility for policies and projects emanating from that ministry. Regardless of the institutional model within which they operate, they are often under-trained and over-worked.

However, because focal points are the primary location for the implementation of a mainstreaming framework, the list of constraints that they experience in the daily execution of their responsibilities serve as a good barometer of the regional climate toward gender mainstreaming. It is disturbing, but not surprising, that their working conditions highlight obstacles at every conceivable stage of gender mainstreaming. These regional constraints include conceptual limitations, difficulties in building intra-institutional coalitions, and a lack of political will and economic resources; each of these plagues the process of gender-justice

64 *Feminist Advocacy and Gender Equity in the Anglophone Caribbean*

mandates throughout the region. Table 3.1, though, shows that even among the focal points themselves, those with primary responsibility for mainstreaming processes face chronic difficulties that would make mainstreaming difficult even if the structural conditions were to change. At the time of this study, few focal points had any specific gender training, a factor that too readily lends itself to reductionist gender analytics and difficulty in shaping mainstreaming objectives that are able to affect the operational and programming details of various purviews or ministries vis-à-vis gender equity.

Table 3.1 Staffing Complement and Gender Training of Regional Units and Bureaus

Country	Existing Staff Complement	Number of Vacancies	Number of Established Posts	Levels of Gender-based Training/Qualifications
Anguilla	4	1	2	One individual with training, level not identified
Barbados	7	1	4	No formal training
Belize	17	1	14	No formal training
British Virgin Islands	3	1	3	One staff member with related training
Cuba	—	—	—	
Dominica	7	1	2	Two: Includes a Certificate in Gender and Development Studies and Msc. Level courses
Jamaica	27	7	26	Three: Levels not indicated
St. Kitts and Nevis	6	1	6	Two: One formally qualified, one with workshop-related training
St. Lucia	7	2	6	Two: Training and qualification at Bachelor's Degree level
St. Vincent and the Grenadines	3	0	3	Two: Gender-related training and UWIDITE Gender Training (latter ongoing)

1. Posts that are formally incorporated as Civil Service Appointments.
2. This figure reflects eight clerical/ancillary positions.
3. Executive Director is a contractual appointment. It must also be noted that Jamaica's bureau has historically been structured with lower wages than other agencies that carry out similar functions.

Despite the establishment of the requisite gender bureaus and desks throughout the region, government budgetary allocations and subventions remain inadequate for the day-to-day operations of these units. Therefore, the effective operation of many of the region's bureaus and desks relies heavily on funds secured from external sources and international donor agencies. Each of the country representatives interviewed indicated that their work agenda depended on support from bi- and multilateral funding agencies.[10]

Despite the invaluable role that such funding plays in keeping bureaus afloat, it may also undermine the sustainability of mainstreaming processes within the region. External funding may potentially lead to managerial complacency on the part of high-level administrators and further be exploited as an opportunity to abdicate their fiscal responsibility for a mainstreaming agenda. External funds may unduly influence the unit's work-plan and vision and exacerbate the pre-disposition to see mainstreaming as a foreign imposition.

Nuket Kardam adds to this funding dilemma by pointing to the negative effects that cyclical rounds of grant writing has on programmatic implementation. A heavy dependence on external funding sources potentially compromises implementation as bureaus must constantly search for funding opportunities as a means of staving off the loss of funds at the end of grant periods, invariably detracting from time and energy spent on programmatic implementation and agenda setting (2007, 98–99). Further, drawing attention to the sources of external funding from bi-lateral agencies and U.S.-based lobby groups has proved to be a very effective strategy that those who are hostile to an equity agenda use repeatedly in their attempt to color gender justice as alien, Northern, and white.

Additionally, the competitive nature of external grants intensifies existing hierarchies between state representatives and feminist NGOs and further fractures the political coherence of the feminist lobby. Nationally and, at times, regionally, competitive grants favor organizations that already have skilled grant writers within their organization, pitting one faction of the women's movement against another, further fracturing the feminist lobby. Difficulties notwithstanding, there are occasional glimmers of success. In 2004, gender bureau heads identified the following among their markers of success:

Barbados

- Establishment of a domestic violence shelter in conjunction with the Business and Professional Women's Club of Barbados.

Belize

- Collaboration with the Ministry of Finance toward the establishment of the Gender Budget Initiative.
- Passage of Sexual Harassment Act and approval of Sexual and Reproductive Health Policy.

Cuba

- 35.9 percent female Parliamentary representation, surpassed only by countries in which a quota system exists.

Dominica

- Collaboration on Organization of Eastern Caribbean States (OECS) Family Law and Domestic Violence Reform Initiative.
- Public education and sensitization through two radio programs: "Talking Gender in Dominica" and "Women's Magazine."

Jamaica

- Allocation of Senior Policy Analyst.
- Increase in total staffing allocation.

St. Kitts and Nevis

- Use of sex-disaggregated data used in Gender Budget Initiatives, Poverty Assessment Surveys, and Labour Market Surveys.

St. Lucia

- Legal Reform: Equality of Opportunity and Treatment Employment and Occupation Act (2000).
- Repeal of Agricultural Workers Act (1979) that had allowed for higher wages for men in certain agricultural activities.
- Criminal Code (2003), reform which decriminalizes termination of pregnancies in *certain* circumstances.

(UNIFEM/Rowley 2003, 15)

The above achievements include cultural, political, and legislative markers. However, these achievements are not without complications and contradictions. Of the points listed above, the St. Lucia Criminal Code (2003) on women's reproductive options is one such egregious contradiction, which I shall revisit in Chapter 4.

Of course, this must suggest to us that as long as the present bureaucratic structure and climate persist, there will be little political will or structural capacity to lobby and advocate for gender change on the inside of the state. Still, such obstacles ought not to categorically suggest the demise of state-sponsored feminism (Barriteau 2003). Rather, envisioning viable alternatives for advocacy on women's issues becomes more urgent. Regionally, Gender Affairs Divisions have been institutionalized with overwhelming impossibilities as their inheritance; consequently, we do not know yet what can be achieved through effective gender mainstreaming in the region.

Sonja Harris has also questioned the extent to which gender bureaus can feasibly exist within the present bureaucratic climate. She argues:

> ... what is required for Women in Development of Gender or Gender and Development cannot automatically be met within the policy framework and praxis currently being pursued. This praxis is characteristically authoritarian, top-down, patriarchal, market driven and dependent on benefits trickling down to the poor (an approach long discredited in the sub-region for its poor results) whereas development planning and programming are by nature participatory, people-centered and focused on equity and sustainability. (2003, 184)

Harris suggests here the need for the region's bureaus to override the institutional pull of authoritarianism in favor of participatory democracy. However, the political rhetoric around "participatory democracy" is such that "participatory demagoguery" is often lauded as "participatory democracy." This is to suggest a distinction between the appearance of popular participation only as sanctioned by the voice of the maximum leader versus rethinking our organizational structures and bureaucratic cultures to encourage broad-based leadership and community involvement in decision-making.

The wide range of responsibilities allocated to implementing units places an unusually varied, multidisciplinary staffing demand on those with responsibility for mainstreaming efforts. Ideally, mainstreaming processes benefit most from having a contingent of disciplinary-based specialists housed within implementing units. The model that I am pointing to is one that will have a range of area-related gender specialists (e.g., economic, legal, and agricultural) housed *within* implementing units to take the lead in sensitizing the respective ministries within the national machinery. Nevertheless, understaffing and a lack of technical skills remain among the most crippling aspects of the advocacy units, making such an approach unlikely. For some, staffing issues were dire, in need of the most rudimentary of staff, such as coordinators or secretarial staff. When asked about their ideal staff complement, bureau heads identified the need for a remarkably varied slate of skills, inclusive of lawyers, counselors, researchers, and project managers.

Within advocacy arenas, apologists tend to explain away the constraints and obstacles that I have outlined here as an intractable lack of "political will." However, as Goetz rightly observes, the "lack of political will" as explanation fails to explain how gender differences and male privileges are institutionalized and normalized within the daily operations of state management (1997, 70). I want to shift analytical scales yet again in order to provide a more nuanced reading of both the performance and the effects of what has been popularly rendered as a "lack of political will." I think it particularly important to pay attention to how heteropatriarchy maintains its privilege within state machinery even as it points to its rhetorical support for gender equity agendas. I also draw on qualitative data to foreground the effect that heteropatriarchal state privilege has on the psyche and emotional well-being (or lack thereof) of the focal points with responsibility for mainstreaming.

Feminists have historically critiqued the ways in which women have been excluded from state resources. I argue that it is now equally as relevant, if not more so, to examine the *terms of inclusion* for potentially discriminatory practices. These kinds of conversations are of critical importance because regional gender bureaus remain the lead change agents within the national machinery and, as such, their legitimacy must be premised upon their capacity to envision and implement the terms of a twenty-first century gender-just Caribbean society and build coalitions of dialogue around this vision.

DEVELOPING GENDER-BASED INDICATORS OF MODERNITY

One of the major accomplishments of the United Nations *Human Development Report* (HDR) has been the introduction of gender-based indicators. Published annually, the HDR ranks countries within the global community based on their performance in three primary areas: life expectancy, educational attainment, and real gross domestic product (GDP) per capita; these three indicators collectively form the Human Development Index (HDI). Beyond its highlighting a country's performance in these key areas, the HDR's ranking is now an important part of ascribing and assessing the legitimacy of state performance in the "modern world." The nationalized desire for "development" and "developed country status" that now pervades the electoral and policy rhetoric of the Anglophone Caribbean[11] is due to the disciplinary links that have been forged between the indicators of "development" and "modernity." An example is provided by Kardam who, in her examination of mainstreaming processes in Turkey, observes that the pull toward the institutionalization of gender equity is due, in no small part, to governments being "pressured in international forums and to avoid embarrassment . . ." (2007, 99). This embarrassment is linked to the increasing association made between gender equity and narratives and indicators of modernity in the global arena.

The HDR expanded its ranking criteria to account for gender-based inequity through the addition of the Gender-related Development Index (GDI) and the Gender Empowerment Measure (GEM). Instituted in 1995, the GDI assesses the gender differentials of the HDI, while the GEM indicates the extent to which women are able to participate in the economic and political life of their country. Despite the fact that these indicators have been established to elucidate the extent of women's access and control of their country's resources, it is undeniable that they also reinstitute a degree of erasure of women's multiple and cultural realities and their localized experiences. The GDI, for example, by maintaining GDP as its measure of economic activity, masks the level of mobility that many Caribbean women have been able to achieve as a result of aggressive income-generating activity as micro-entrepreneurs, or "higglers," within their informal economies. Of course, the corollary of this is that we are similarly unable to address the forms of economic, physical, and sexual exploitation that occur with women within the informal economy, rebounding nonetheless to increased profit margins within the household and formal economic activity (e.g., tourism based sex work). Similarly, it masks the difficulties experienced by these very women in the informal economy as they approach formalized institutions, such as banks and state-based micro-enterprise lending agencies. Further, the composite indices of the HDI (longevity, GDP, and educational attainment) reflect the capacity of already financially endowed states to perform well in the provision of resources to their citizens.

Nonetheless, the usefulness of these indicators is evident when one accounts for the gender differentials of development. By 1997, two years after the implementation of the GDI and the GEM, countries were increasingly made aware that economic stability did not necessarily translate into similar levels of stability for women. After adjusting the HDI rankings for their responsiveness to the gender differentials of access instructive disparities emerge between a country's economic buoyancy and women's ability to benefit from this buoyancy. An overview of the GDI-adjusted ranking of countries in 1997 reflects the economic disparity and inequity women experience globally in controlling the most basic benefits that can be derived from their country's "development." Of the sixty-four countries ranked in the "High" cohort, 25 (39 percent) dropped when adjusted for the GDI, whereas all of the forty-four countries ranked in the "Low" cohort of the HDI improved in their GDI.

Over its life, the GDI has effectively shown that gender equity is not only a function of economic stability. Table 3.2's comparative discussion highlights elements of this disparity and further suggests that addressing gender inequity is not purely a function of economic prowess but an explicit commitment to addressing the terms of inequity. Nevertheless, because the GDI is already based on the economic stability of a country and its capacity to provide socio-economic opportunities, in poorer territories even where country rankings may have improved, women's absolute level of access is precarious and threatened by the overall predicament of scarcity. An improved ranking, therefore, cannot categorically suggest acceptable standards of well-being and livelihood for women.

The GEM is also enhanced by a similar qualitative and contextual critique. Women's access to the formal reins of decision making is imperative to the long-term goal of achieving gender equity. Interestingly, many of the region's territories improve in their ranking once the GEM is taken

Table 3.2   Comparative HDI, GDI, and GEM Ranking for Selected Countries[1]

Country	1997			2000			2007		
	HDI	GDI	GEM	HDI	GDI	GEM	HDI	GDI	GEM
Barbados	25	17	18	25	17	14	31	30	30
Grenada	54	—	—	54	—	—	82	—	—
Jamaica	83	63	—	83	63	—	101	89	—
Trinidad and Tobago	40	32	21	40	32	17	59	55	23
Netherlands	6	11	6	6	11	10	9	6	6
Switzerland	16	20	13	16	20	12	7	9	27
Sweden	10	3	3	10	3	2	6	5	2
United States	4	5	11	4	5	7	12	16	15

[1] Table compiled drawing on data from the respective Human Development Reports.

into consideration. This improvement, while noteworthy, prompts some discussion of the indicator's weakness in relation to regional gender justice. While GEM's indicators highlight the importance of women's access to formalized political activity, GEM simultaneously reinforces a mainstream private/public dichotomy of what kinds of activity constitute legitimate political activity. The intersectionality of women's lives consistently defies this binary; as such, much of the important work done by community activists remains erased, invisible, and, at best, undervalued.

The GEM's inability to challenge formal definitions of "political activity" consequently cannot account for the work that women do within the less formalized arenas of political activities. Moreover, Caribbean political parties have historically exploited the canvassing done by their female supporters. This activity arguably improves Caribbean women's status in their community as community-based activists and their roles as mediators between their communities and prospective government agents and state functionaries. However, while Caribbean women remain the primary actors in this regard, they have yet to become the benefactors of increased levels of access to formalized corridors of power (Figueroa 2003). The GEM's replication of the informal/formal dichotomy of political activity in no way elucidates the disparities that arise between these spheres, nor does it validate the connections. The GEM thus erases the importance of community political activism as a distinct political product within women's domain of activity. Further, the GEM in its present configuration is in no way able to measure the extent to which women's participation in the formal political life of their country translates into meaningful political advances for the women in that country. As such, the Caribbean stands as a textbook example of an improved GEM ranking without commensurate substantive gender equity gains as a result of such participation.

Despite the weakness of these indices, they have initiated a new and critical lexicon on the terms of "development." The global discourse on women's rights within the worlds' nation-states is now informed and governed by, among others, four UN-sponsored International Women's Conferences, the CEDAW[12], now ratified by approximately 170 of the world's countries, and the Beijing Platform of Action, which outlines a number of activities to be achieved in twelve different strategic areas of change. The terms and content of these internationally established mechanisms of monitoring and evaluation assess the efficacy with which nation-states, themselves unequal in power, have actually implemented gender-sensitive legislation and a bureaucratic structure that facilitates the practice of gender equity.

These indicators have initiated a way of inscribing national identity that further incorporates and institutionalizes women's bodies into a discussion on modernity and re-invokes the historically drawn lines between the "West and the Rest."[13] For a country in the supposed "developing South," there is a shifting yardstick that marks its status as "modern" by virtue of its ability to speak about the terms by which its female citizens exist. Therefore, nation-states in the Anglophone Caribbean increasingly tend to craft themselves (if only rhetorically) as gender sensitive in their planning approaches. The region's state-managers are invested in representing themselves as gender aware to the world community and are indeed called upon to do so through annual ranking and periodic reporting to conventions ratified, as explicitly stated in the *National Report on the Status of Women* (1995):

> The Government has ratified several conventions that impact in a general way on the welfare of women ... In addition, the Government in 1990 ratified the Convention on the Elimination of all Forms of Discrimination against Women ... In the past twenty-five years, conscious efforts have been made to repeal or amend all known discriminatory legal impediments to the advancement of women. (43–44)

These gender-related factors have become critical components of how contemporary nation-states attempt to craft an identity of international merit. In this light, territories in the Anglophone Caribbean, like many other displaced and disadvantaged actors on the international scene, have found themselves caught in a race of ratification. This is a race, however, that must be identified for its hollowness and incapacity for operationalization primarily because it facilitates the political rhetoric of gender sensitivity and inclusion without commitment to the infrastructural, ideological, legislative, and financial implications of what it means to be a signatory to conventions that are aimed at facilitating gender equity. As such, a huge dissonance emerges when one tracks the strength of the key gender focal point within the national machinery against the ratification enthusiasm. For the same period within which many of the region's governments would have ratified documents such as CEDAW, Nairobi Forward Looking Strategies, or the Beijing Platform for Action, the institutional status of women in many Caribbean territories experienced several transitions from Desk to Bureau to Division. Yet, insofar as one can identify these organizations as the barometer of the state's thinking on women and gender equity, it is particularly telling that these organizations have been notoriously underfunded, limited in ministerial influence, and poorly staffed both in terms of personnel and training. These things are now well known and cannot be disputed.

However, I am more concerned with the subtleties of incapacity. That is to say, even in situations in which there may have been increased financial allocations or an increase in staffing and technical resources, what are the qualitative and unmapped elements that debilitate the extent to which the region's bureaus can effectively realize any goal of gender equity? To what extent can one conclude that the region's bureaus have been institutionalized in ways that predispose them to failure?

DE-CENTERING WOMEN: PLANNING AND THE GENDER AFFAIRS DIVISION OF TRINIDAD AND TOBAGO

Looking toward women's experiences has offered feminists a way of making visible subjugated subjects, redressing their historical marginalization, and offering insight into how discourses come to reside in a subject's understanding of herself (Scott 1991; Stone-Mediatore 1998). Having examined the global and regional context as well as the international pull of nation states toward the UN rhetoric of modernity through gender equity, I would now like to center the experience of actual femocrats as they work to navigate these three tiers of mainstreaming realities.

In Trinidad and Tobago, forty years after independence, it can be said with some certainty that a national agenda for women's equality has been an eclectic, if not an ad hoc and reactionary approach. The interim forty years have been characterized by a number of political and economic upheavals. The emergence of new political actors[14], the 1970s Black Power movement, an oil boom and crisis marking both economic prosperity and depression, IMF (International Monetary Fund)–imposed structural adjustment policies, an attempted coup d'état, and peaceful elections in the midst of ethnic diversity and rapid change have been some of the defining features of Trinidad and Tobago's first forty years as a nation-state. This overview also gives a partial outline of the fraught political climate in which the question of women's equality would be placed on the national landscape. In this context, the issues that would have been prioritized within the state machinery could be categorized as those seen to have international impact, those promoted by strong advocacy of a local feminist lobby, and/or changes in the voting demographics of any given period. The imperative to address gender equality in Trinidad and Tobago, therefore, emerges at the intersection between the international mandate and local feminist lobby within the nation-state.

Prior to the declaration of the International Women's Year in 1975, any approximation to women's issues in Trinidad and Tobago fell largely under the purview of partisan political campaigning by the women's arm of their perspective political parties or under the framework of "generic" legislative framework of services provided to women and men and through the dispensation of welfare services and the formulation of a social safety net. As discussed in the prior chapter, predominantly male nationalist leaders of the post-independence period adhered to a welfare approach in the provision of social benefits, albeit void of concerns for women's empowerment and gender equality. Therefore, while this welfare approach laid the groundwork for the development of a social consciousness and a sense of entitlement in the areas of access to health care, education, and basic needs and amenities, it did very little to move women beyond their traditional roles as wives and mothers.

The UN Decade for Women, 1975–1985, was a pivotal period in turning global attention to the institutionalization of women's issues. This period added momentum and synergy to an already vibrant women's movement in Trinidad and Tobago, which had crystallized into a number of pressure groups focused on advocacy for women's rights. The language employed by many of these "second wave" activists in this decade reflected the socialist-based politics of the Left and a catalytic blend of issues, such as sexual autonomy and control of women's bodies (e.g., Rape Crisis Society) and social equity (Caribbean Association for Feminist Research and Action, 1985, and Women Working for Social Progress, 1985). During this period, organizations such as the Housewife Association of Trinidad and Tobago (1975) called the state to a degree of accountability on issues that addressed women's domestic and community management responsibilities (Reddock 1998). State officials, therefore, had to up the ante in terms of their response to these organizational demands in ways that did not allow any of the red-herring comfort that state-managers would later find in the ratification race. Despite the historical force of a strong women's lobby, several structural, ideological, and conceptual flaws remained. These weaknesses rendered state policy and planning responses as less than credible and, one might say, deliberately ineffectual.

For Trinidad and Tobago, several legislative milestones with direct impact on the advancement of women, if taken at face value, make a convincing argument that the state has been proactive in its efforts to improve the status of women. Among some of the milestones have been the Domestic Violence Act, No. 22 of 1999 which repealed the earlier Act of 1991, improving on the compensatory and protective components for victims. The Maternity Protection Act, No. 4 of 1998 guaranteed paid maternity leave for 13 weeks and protection against dismissal as a result of pregnancy. The Cohabitational Relations Act, No. 30 of 1998 accords women who have lived in a common-law relationship for five years or more with the right to apply for an adjustment of property and child maintenance upon separation (*Initial, Second and Third Periodic Report for Trinidad and Tobago* 2000). Further, Trinidad is one of the few Caribbean countries that has submitted periodic reports on the CEDAW convention, albeit approximately ten years after having initially ratified the convention.

The Gender Affairs Division is the only state-based institution explicitly mandated to address and facilitate the holistic implementation of issues related to Trinidadian women's empowerment and gender-based inequality. In keeping with their mandate to effectively "promote Gender Equity and Gender Equality through the process of Gender Mainstreaming in all Government Policies, Programmes and Projects," the Division is designated and perceived as the lead agency in the promotion and monitoring of measures that facilitate women's advancement.

The Trinidadian Gender Affairs Division is expected to provide and receive support at two levels, through intra-ministerial networking and through the resources and capacity built into the Division itself. However, on both counts the state's treatment of the Division mirrored the marginalized status of women in Trinidad generally. I want to consider how the capacity of the region's Gender Bureaus to actually realize their mandate was constrained by what I see as three important components to the production of incapacity: a struggle against suspicions of "illegality," constructed narratives of incompetence, and a colonially driven bureaucratic structure generally hostile to change and obstructively hostile to gender equity. I have already addressed the colonial structuring of gender relations within bureaucratic formations in Chapter 2. I want to continue this institutional critique by examining how tropes of illegality and incompetence are deployed against gender focal points.

From its very inception, the Gender Affairs Division struggled against a sense of "illegality," as evident in the following narrative given by one of the project officers who would have been among the first complement of staff within the Gender Affairs Division:

> *P.:* . . . so they hired us, they hired us ostensibly to implement the Inter-American Development Bank (IDB) plan of action which was agreed to by the government. When we got there, there was politics . . . . the policy changed . . . . In fact, when we got there the P.S. [Permanent Secretary] said we were illegal, there shouldn't be a division for women alone, so he refused to deal with us. That was [name blurred on tape]. I will never forget that name. So they used to give trouble about our contracts and payments.
>
> *Rowley:* But illegal in what sense?
>
> *P.:* He said he couldn't understand how the government could have a division to implement programs for women . . . the Women's

Affairs Division. We were leaving out one half of the population. The usual . . . .
Rowley: Nonsensical argument.
P.: Yeah, women's affairs.
(interview with member of the first staffing cohort March 18, 2002)

The term "illegality" repeated itself throughout my interviews with the state's functionaries. This reference to illegality reflects the stranglehold that additive renderings of gender have on what is seen to be appropriate gender planning. "Illegality" thus became a way of describing a conceptual location that did not define gender relations as a categorical attendance to the concerns of "men and women." Any other organizational model to this rudimentary understanding of gender, regardless of its empirical and conceptual merit, was seen as "illegal," and, therefore, not a viable contender for accessing state resources. This lack of perceived legitimacy by state-managers subsequently resulted in deliberate acts of sabotage through tardy salary payments and contractual instability.

Populist assumptions about "gender" as categorically requiring attention to men, when enacted as the basis of policy and decision making, opens hitherto untapped resources previously unavailable for women. This depoliticized terminology also makes it more difficult for feminist activists to argue for women's socio-economic, physical, and political vulnerabilities. The following dialogue between myself and a former Minister of Government with responsibility for gender affairs reflects the insidious hold that rudimentary "gender" models have produced on the inside of the Gender Affairs Division and the state generally:

Minister: I remember I was going to a meeting in India, I think in 1999, and there were a number of issues that we were going to discuss in India and there was opposition to my going to India. But I said what are the things we are going to do and when I talked about the males, because we were going to deal with that thesis on male marginalization. I think it is Miller . . .
Rowley: Yes, Errol Miller in Jamaica.
Minister: . . . When I mentioned that, in fact not only did I have to mention it, I had to bring documentary evidence that the male marginalization thesis was being addressed in India . . . that allowed me to go. (interview with former Minister of Gender Affairs March 10, 2002)

This exchange reminds us that women's political success cannot merely be numerical access to the corridors of power without a commensurate concern for redressing the patriarchal institutional cultures and procedures that exist within these corridors. Here, the Division's work agenda was governed by the need to respond, negatively or positively, to the growing concerns about masculinity in the region. The growing correlation between masculinity and marginality successfully redefined and undermined the terms against which decisions were made for women's empowerment as well as the levels of access that women had to state resources.

However, the term "illegality" substantively extends beyond the Division's ability to procure the resources to facilitate administrative expenses or work plans toward gender equality. In addition to tropes of "illegality," state-managers also deploy feminized narratives of incompetence and madness, producing an even greater sense of "un-belonging" and organizational malaise by gender focal points. The effects of these stratagems were poignantly expressed, as in the following dialogue given by one of the Division's former directors:

> *Director:* . . . I was a mad woman anyway and this was not an NGO, this was what people would have perceived. But the key would have been to operate like an NGO inside there because of the politics of the thing, knowing fully well that you wear your masks at the critical time.
> *Rowley:* I want to pick up on something you said somewhat facetiously, "I'm a mad woman anyway" . . . What makes you use that idea?
> *Director:* [giggles] Well, I said it in a pleasant way, but there were lots of feelings. I did not want to manage the Division the way the public service and the Minister wanted me to manage it. I absolutely did not want to do it! My training in feminist methodology didn't allow it. . . . .I disliked hierarchy to the bottom of my heart. But my Minister kept shouting at me to direct! "Why don't you direct?!" . . . I understood power well, so what I quickly had to learn to do was to wear all these different masks. So for some people on staff I cyar [can't] manage at all! And for me I'm saying I don't want to manage because I don't want to do it in the same way as the teachers do, as the community officers do it, the nurses do it and so on. Take off your shoes, let us go down in the market street, let us talk with the women on Charlotte St. [becoming more and more animated]. You understand that's how we have to do it. We can't look pretty and do that. You understand, so I had to be mad . . . [holds head in exasperation] so you just give up. (interview with former Director of Gender Affairs Division April 5, 2002)

The face-off between feminist discourse and state processes leaves casualties in its wake, foremost among which are gender focal points. The personal and psychological articulations that punctuate the progression of the director's narrative are striking. In this exchange, fatigue, a sense of defeat, and the heft of having one's integrity assaulted through narratives of madness all mark the result of hostile bureaucratic responses to feminist challenges.

Bureaucratic cultures, by their own design, produce dispassionate, distant similarity of procedure and function across a network of ministries. This culture is an oppressive force for a division such as the Gender Affairs Division, which ideologically and functionally has been designated to produce change within its counterpart agencies. Feminist practitioners within bureaucratic structures are constantly required to engage, woo, and persuade disinterested bureaucrats whose professional disassociation blinds them to the ways in which, as individuals, they are simultaneously part of the social order and may potentially benefit if changes are successfully implemented. In the absence of such identifications, feminists are rendered as alien in daily work environments. Straddling this insider/outsider relationship within the national machinery produces psychologically incarcerated change agents torn between the desire for gender-just change and the predetermined organizational structure that resists such change.

The reference to "madness" is not accidental, even though the director herself dismisses her statement as conversational filler by the use of the adjunct "anyway" ("I was a mad woman *anyway*"). Mental instability and irrationality have a long philosophical history as the basis of women's exclusion from public life (Pateman 1992). Madness, even if used facetiously, opens up several avenues for sabotage due to supposed or imputed irrationality and narratives of incompetence. Ascribing madness to any change agent, inclusive of those pursuing gender concerns, sets the stage for dismissing the speaker's point of view and perspective. Madness, then, becomes both a silencing strategy as well as an indicator of how bureaucratic institutional cultures produce psychological discomfort for feminist change agents. In addition to ascribing questionable narratives to the capacity of change agents, another important aspect of weakening the bureaus globally has been the systematic detachment of the Gender Affairs Division from a community of feminist activists or the further instrumentalization of this community toward the furtherance of state goals rather than the feminists' interests and intent of their varying constituencies (von Braunmühl 2002, 73).

Feminist practitioners may experience their position within state machinery in even more isolating terms if the members of the larger feminist civil society view them as having capitulated to state forces. Capitulation notwithstanding, mainstreaming, told from the perspective of feminist practitioners within state machinery, is rarely a celebratory story. The increasing institutionalization of matters pertaining to gender makes it difficult to avoid engagement with state forces. However, those who do attempt to engage may be labeled somewhat derogatorily as "femocrats," feminist technocrats who benefit financially from huckstering in a brand of "state feminism." At best, they may be seen as a necessary but suspect rung in the climb to legislative and political advancement for women (Gouws 1996).

In the above exchange, the former Director, in reflecting on her tenure with the Division, highlights the need to claim "NGO status" while located within the bureaucratic framework of the state. The NGO analogy I find to be an apt one in the ways it asserts the need to walk the sometimes blurred line of an insider/outsider identity within the state machinery. Metaphorically, it also calls forth an oppositional critique as evidenced by attempts to draw on feminist methodology and non-hierarchal managerial styles while cranking the levers of state access. This double play of interaction is even more critical in light of my argument in this chapter that one of the challenges for gender focal points is the need to clarify and strategize around their terms of inclusion. This critical stance increases the likelihood that gender focal points will be able to name and mark the silencing stratagems at work within their respective agencies. When focal points challenge the practices that attempt to minimize them and trivialize their agendas, they simultaneously engage in the creation of a more gender-just society. Such a challenge stands at the core of an oppositional critique.

GENDERIZING THE PLANNING DEBATE: CONCEPTUAL AND PROGRAMMATIC AMBIGUITIES WITHIN TRINIDAD'S GENDER AFFAIRS DIVISION

Conceptual ambivalence produces programmatic hesitancy and inefficient use of resources. The principal area of ambiguity with regard to bureaus regionally is the "gender dilemma" that must be confronted at the most rudimentary level of "gender mainstreaming." The dilemma around "woman"/"gender" in the day-to-day running of the Division's work produces a great deal more than semantic ambiguity. Rather, this dilemma supports the narrative of women's "illegality" within national machinery and secures the propensity for gender-planning agencies to become a safe place in which heteropatriarchy can hide.

Guided by the understanding that various categories of women and men in any given society exist both differentially and hierarchically, gender mainstreaming prompts us to think critically about the ways in which any policy, plan of action, legislation, or program will impact the livelihoods, well-being, and capabilities of women and men.[15] Gender mainstreaming first requires clear and explicit assessment of the location, condition, needs, and demands of the various constituencies (e.g., not merely by sex, but by race, caste, sexuality, and other modes of inscribing inequity) who will be impacted by the proposed plan of action. It further insists that state functionaries with responsibility for gender mainstreaming, having gathered this data, incorporate such as an integral element of both the substance (e.g., legislation, policies, and projects) and form (e.g., organizational framework and practices) of planning approaches (Kabeer 2003). However, the "gender dilemma" as it exists regionally in the Anglophone Caribbean is such that these factors are not consistently incorporated into the implementing frame of gender mainstreaming. In the absence of a clear conceptualization of gender, there is no sense of which categories of individuals should be addressed or how resources should be allocated by already fiscally strapped states. This conceptual ambiguity, in turn, compromises the legitimacy of the claims made by women as well as the politics and purpose that inform the Gender Affairs Division's work program.

In the Caribbean, the shift from "woman" to "gender" marks the resurgent power of patriarchal impulses in state formation. The latent fear of "female ascendancy" is an anxiety that has prompted state managers to embrace the rhetoric of "gender" in ways that facilitates the state's unwillingness to engage issues that are directly related to women's vulnerability. The danger with the use of the term "gender," as Caribbean feminist scholars often remind us, is that it has been used by state officials to uncritically and categorically replace the constituency "women" with a more nebulous rendering that is at best a social constructionist approach and at worst a crass reductionist approach of gender as "men and women" (Barriteau 2000).

This "gender dilemma" took an interesting turn in Trinidad and Tobago in 1997 when the change in nomenclature from "woman" to "gender" resulted in an impasse between members of the feminist NGO community and the state-based advocates of the change. This change was more than a nominal one; it generated conceptual uncertainty among staff, increased the levels of ambiguity about the Division's target group and the legitimacy of resource allocations to these various constituencies. This, of course, had implications for the work program that was pursued by the Division.

The transition brought the "gender dilemma" to the fore. The conceptual ambiguity among staff was palpable. One of the main points of conceptual uncertainty in the most fundamental sense surrounded what the term "gender" meant and, by extension, how did it then modify the Division's constituency and work program. The transition was not accompanied by training or re-education, and in the daily running of their Division, the staff was called on to operationalize a term that had little or no meaning to them. A range of different (mis)understandings came to the fore when I asked the Division's functionaries how they felt about the Division's transition from woman to gender:

> I don't think the Minister understood conceptually what was the difference between women and development and gender and development. Or if she understood, she didn't really care. She just decided, you know, to attract men and get away from some of the negative vibes of women's affairs, let's do gender!

> It was women's affairs and then became gender affairs. She [the Minister] felt that in working for women you didn't use a gendered approach if men weren't targeted with gender training and gender information . . . This happened a lot because of domestic violence and because a lot of policy makers were men, and also because the Commonwealth [Secretariat] was also changing their approach from girl and woman to gender . . . A lot of people in the women's movement felt that we hadn't achieved the goals of Beijing . . . we hadn't arrived. They felt that we were giving ground.

Well, early we shifted from women to gender and I understood the gender. As a matter of fact, I had bought into the shift in the concept because you can't help women in isolation, *unless you talk about the men too*. I had a little, um, in terms of my version of what I saw as gender, got a little, um, a little opposition within the Gender Affairs Division. Because from the literature, even though the literature was saying gender, you got this feeling that it was saying women. (emphasis mine)

I just thought it was so stupid because to me, if you want to do it, you're going to ask what informs it. So when gender, as I say, exploded, and it has exploded on the country . . . this is why we have all this conceptual unclarity and confusion. (interviews with first cohort of staff members March 2–4, 2002)

These responses provide direct insight into the various arguments and positions that emerged around the shift. The marginalizing veto power of state power was so overwhelming that despite the dissent or disapproval of these change agents, there was a sustained and attendant narrative of disempowerment and inability to initiate change. The operationalization of the term "gender" reflected the super-ordinate position of the Minister's mandate over and above the Division's technical expertise. There was scant consideration of what the change would mean in a bureaucratic climate governed by male marginalization rhetoric. Neither was there a sense that "gender" was required to work in tandem with "mainstreaming"; nor was there yet to be a discussion on the relationship that should exist between substance and form.

These conceptual deficiencies neutralized the empowerment and advancement offensive that inhabited the spirit of CEDAW and, in addition, stunted the possibilities that could result from a radical use of the word gender. One of the major by-products of this transition was a heightened "sensitivity" to the needs of males in society. Regardless of the privilege and statistical and socio-political dominance that Caribbean men experience generally, crafting a male programmatic agenda as an additive gained paramount importance. This anxiety around masculinity resulted in the creation of a Male Support Unit. According to the Minister who spearheaded the transition, changes were necessary for her Ministry to respond appropriately to the perceived sense of male insecurity, as expressed here:

Rowley: How would you assess the national climate on women's issues?
Minister: There are two elements I think that I caught. One is because of the concerted effort we used to address women's issues. We find the men have been feeling more insecure and to some extent marginalized and they have expressed this. I told you we had a male support group and now a unit, um, so we've had to deal with the male backlash if that is the

> correct word, um, because of the, we've even been accused of, of strengthening the woman as opposed to the man and therefore making men more insecure. (interview with former Minister of Gender Affairs March 10, 2002)

The complete disconnect between the question and the Minister's response reflects exactly where the national climate is on women's issues: locked within, and submerged by, a growing anxiety about masculinity. The thesis of male marginality has been on the ascendancy *as part of* the political rhetoric of gender. Both the currency and danger of this rhetoric are not to be underestimated. The male marginalization model propagates the popular belief that men and women exist in an obverse relationship with each other, that is to say, acknowledging the specificities of one simultaneously means that one ignores the demands or needs of the other (Barriteau 2002). Apart from the simplicity of this model, the danger of incorporating these ideas into planning and policy formulation invites a planning model that disregards the socio-cultural privilege that Caribbean men hold. It distracts us from understanding the ways in which state institutions can be best poised to assist Caribbean men, that is to say, assessing contextualized need in relation to the quest for equity. Finally, it compromises the certainty with which women and ministers with perceived responsibility for women are able to hold the state accountable to them as a constituency.[16]

Conceptual ambiguity around the "gender dilemma" is one of the ways that we see policy heteroglossia at work within gender bureaus and other national machinery. In the policy heteroglossia of gender, we encounter a struggle between the rhetorical devices, policy aims, and objectives of various approaches to gender mainstreaming such that one becomes "... a centralizing (unifying) tendency, the other a decentralizing tendency ..." resulting in a stratifying effect within institutional discourses (Bakhtin 1982, 67). In the mainstreaming of gender within the region, the official rhetorical embrace of practices and strategies reflects an additive understanding of gender. Within *this same framework,* we find the increasing marginalization of political practices and strategies that pertain directly to women as a marginalized constituency, or alternatively, we find resistance to any embrace of gender that attempts to reckon with the implications of equity. This results in a stratifying tension and contention between and among these approaches, which manifest in contradictory ways for gender mainstreaming.

What I've referred to here as policy heteroglossia manifests in the programming ambiguity in the implementing unit. What is the appropriate slate of programs to be implemented in the achievement of gender equity is a recurring question within the implementing unit. Skills-training programs feature prominently within the programmatic agenda of many of the region's bureaus, offering a wide range of skills that could potentially gear women for a greater degree of domestic efficiency to participation in atypical, non-traditional vocations. Skills-training programs for "women only" are not always sufficient grounds on which to differentiate the work done by implementing units from that of other mainstream ministries or divisions. This is no more the case than the categorical inclusion of men within programs and projects providing adequate grounds for the use of the term gender. What distinguishes the gender mainstreaming programming from a daily male-oriented activity is the capacity to understand the role that such programming plays in envisioning the structural and programmatic interventions and strategies necessary to counteract inequity.

Gender-related policy heteroglossia as enacted in the Anglophone Caribbean compels gender planners to take an apologetic stance for initiating programs that confront women's statistical marginalization from access and control of resources. The "gender dilemma" weakens aggressive ownership of any programmatic agenda that appears to address the equity-related needs of women. It is, therefore, not surprising but nonetheless worrying that the programmatic activities with obvious capacity towards women's empowerment, for example, non-traditional training programs, have been referred to as "not a gendered program but a *skills* program" by gender focal points. When touted in gender-neutral terms, the programmatic challenges to inequity are rhetorically neutralized.

"The Inter-American Development Bank (IDB) Non-Traditional Skills Training Program" at the center of this interview provided women with atypical skills in areas such as upholstery, masonry, and electrical wiring. The program also incorporated a gender sensitization component for participants and tutors. Originally, the training model was conceptualized to "increase the level of skilled marginalized but substantial sector of the labor force" (TC-94-04-37-1).[17] Yet, hosting a program not explicitly conceptualized to address gender issues within the Gender Affairs Division presented a number of anomalies. The first anomaly arose from women who expected that programs emanating from the Gender Affairs Division would make a significant difference to their lives yet found that entrenched gender discrimination in the wider labor market did not always make it so, a feature exacerbated by the participants' internalized gender socialization. Non-traditional training is of no intrinsic value to women without a vision for the type of output that will enhance women's capabilities not as wives, not as mothers, but as self-sustaining agents. Without a vision for change, gender focal points fall into the trap of conceiving change via numerical indicators, that is to say, how many projects and how many women.

Numerical indicators serve an important role in the Division's evaluation process. However, the effects of focusing on numerical indicators do not end at the institutional level; they can also affect the women who interface with the Division, who in turn fall into a cycle of certification. In this cycle, participants move from one skills training program to another, increasing the number of participants in the programs offered by a division and expanding on the skills without a clear sense of how these skills actually make a difference to the quality of their lives or their ability to challenge their lived practices of inequity.

This lack of clear application was immediately evident in the following respondent's narrative. Joan was unemployed with five children. She moved back into her parent's home to care for her ailing mother and benefitted from not having to pay rent but lamented her loss of autonomy. She rationalized her participation in the non-traditional training program offered by the Division as an avenue to "bettering" herself. More in-depth discussion revealed, however, that since graduating from the program, Joan had not been able to use the training given to her. In fact, she had not even thought seriously about how her training in electrical installation could be used:

> *Rowley:* Were you satisfied with the quality of training that you received?
> *Joan:* Although I find they should have gone a step further. This was just the first part; they should have a more advanced part to the program.
> *Rowley:* What about opening avenues for you to make money? How do you see that happening from the skill? Because it's one thing to have the skill, but you have to make the skill work for you.
> *Joan:* Well, I will need, such as, um, well, like, going into business?
> *Rowley:* However. You decide.
> *Joan:* Well, I could go into a business and going into that business is like going into a contractual something and whereas I could have people working . . . now working for me. Now I know, electrical, I wouldn' be able to do it, but I can have a qualified person working for me so that when you do something, I could go and say, well yes, I'm satisfied. (interview with female participant in skills training program May 8, 2002)

When asked about her projected plans to use the training, there was little indication that Joan had even thought of the training in terms of self-maintenance, hence, her delayed response and need for clarification. When pressed, she began to build what sounds more like an entrepreneurial flight of fancy rather than a clearly thought-out business plan, even though the curriculum of the program required her to think through such a business plan. In this exchange, there is no clear sense of what her envisioned supervisory role would require her to provide in terms of start-up capital or technical expertise. When prompted later in the discussion to think about the usefulness of the training, she scaled down her response significantly:

> *Joan:* I went in [to the training program] because I wanted the . . . ah [I] fixing meh [my] place and I would learn more about wiring and ah went in and I enjoyed it. It was good and I was glad in a sense that I did, because at home I'm able, there was a cord, I watch the cord and I say, but wait . . . . it had a socket outlet on to it and I was able to take out the socket outlet, rewire it over, fix it over, take out the plug, cut the wire, put on a good plug, and fix it and so on. (ibid.)

Further discussions with Joan showed that she accumulated a number of training experiences and had yet to make any of them work for her. She had done an upholstery course, an electrical installation course, and was about to embark on counseling—all of them "atypical" skills training—and none of these have changed her material condition.

Similarly, one of the employees of the Gender Affairs Division voiced this concern for the "cycle of certification." In critiquing the Women's Second Chances Program, the agent noted that many of the participants merely accumulated certificates of no general value:

> . . . especially in the agricultural program, they just come in to collect the stipend for three months and then they go back out there, and they really don't have the funds, the infrastructure, they don't have anything to continue the program or to continue anything they have learnt. (interview with staff member of Gender Affairs Division April 8, 2002)

The second anomaly is the paradoxical resistance to name a project resident within the lead focal point for gender-just change as one with a gender focus. Consequently, though housed within the Gender Affairs Division, the training program's gender sensitization module was ironically the most vulnerable element of the program's training content. Dialogue with the NGO agents responsible for the gender component of the program noted that for the duration of the program, the regional program review questioned and threatened the relevance of the gender modules continuously. The Coordinator of the program acknowledged the importance of the gendered component, but when asked if a gender impact assessment had ever been conducted to explore the benefit of these modules to the beneficiaries, she replied:

> I just have the same general sense from talking to the women about how the program would have impacted overall, skills apart. The coming together with women similar to them, um, and we've had women who have been brought into the program being referred to us because of domestic violence in their homes and you hear them now and you know that they understand. Even though the program didn't set out to do that, but they understand that it wasn't their fault. And they understand that, you know, they have a new independence and they don't have to stay . . . . *But you have to remember the objective of the program, so the framers of this not going into that"* (emphasis mine; interview with Coordinator of Skills Training Program April 10, 2002)

The Division's mandate of women's economic empowerment is cognizant of the fact that women's control over money, assets, and livelihood translates into a better standard of life. The gendered component of this assumption, however, is that without adjusting the power relations in the home, the community, and the marketplace, economic empowerment in isolation will not necessarily translate into an improved status for women generally. Returning to the terms of inclusion, it is disconcerting that such an apologist position regarding the need to assess the gendered impact of the project itself came from *within* the Division itself.

Structurally and conceptually, the Gender Affairs Division as the primary indicator of the state's position on women reflects a position of ambivalence and chronic disregard for the terms that inform its own raison d'être. Yet, skill training remains the main indicator of the Division's successes. In the midst of this welfarist, stipend approach to skills training, the only program (IDB Non-Traditional Training Program) to incorporate a gender empowerment component has done so apologetically, and that gender module remains under threat. Assessed against the lives of women such as Joan, who look to these training programs to provide financial stability, greater sense of purpose, and direction as heads of households, it is difficult to clearly point to the unique contributions that the Division makes to their lives.

Many of the Division's former employees felt that the enthusiasm with which the Minister pursued the transition to gender was a strategy of political survival rather than "a thing to really help women." The Minister herself in the following exchange, surprisingly enough and somewhat unwittingly, corroborated this:

> *Rowley:* Again, the question of political will. Referring to your earlier statement about the importance of timing when approaching the cabinet, the importance of understanding what are the topical issues and using that as an opportunity? Do you think that the shift from woman to gender would have been, in a sense, a strategy of approaching Cabinet?
> *Minister:* You bet, you bet! Yes, yes, because the members of Cabinet are male. They largely also hold, in general, the predominant views of men-women relationships. I mean not in the extreme, they won't say you should beat your wife . . . (interview with former Minister of Gender Affairs March 10, 2002)

"Gender" as the basis of planning can be conciliatory and aimed at appeasing, rather than challenging, the assumed and traditional views among different categories of "men-women relationships." Women's issues, if driven by a feminist mandate that challenges the status quo, will hold a precarious and antagonistic opposition within bureaucracies. There is no easy way to lobby the state to address women's issues while residing within the institutional machinery of the state. The transition from woman to gender cannot be dislocated from the Gender Affairs Division's attempt to gain greater institutional credibility via a less confrontational or overtly woman-centered approach. The discord that resulted among different feminist stakeholders, such as the women's movement and various academics, was premised on the sentiment that this transition was not in the interest of improving the status of women and served to create a situation where women's entitlement and access were being eroded.

POLITICAL CORRECTNESS VERSUS POLITICAL WILL: USING WOMEN AS AN ELECTORAL RESOURCE

Throughout this chapter, I have been concerned with the disjuncture between rhetorical practices of gender mainstreaming and the lack of substantive moves toward gender equity. As discussed in Chapter 1, in a societal context in which so many of the activities emerge from a welfarist modus operandi of "doing for" women rather than addressing the more critical question of "how are we creating systemic change for" women, these programs secure a Band-Aid approach to gender equity that shows little or no awareness of the terms upon which one must plan holistically for change.

Many of the programs that emanate from the Gender Affairs Division are primarily adult education programs for women (e.g., adult remedial or vocational), which is a critical component of women's advancement as they address educational imbalances that may confront women. The lack of feminist rationale and focus, however, makes these projects problematic. Throughout the interviews conducted with both past and present functionaries, it became clearer that the need to be seen to be "doing something for women" governed the overt project-oriented work program. A theme that arose with some of the Division's representatives was the importance of portraying a sense of "busy-ness." In reality, this imperative of "busy-ness" or "doing something for women" reflected a number of competing partisan interests. Rather than finding compatibility with the Division's agenda, a range of project-motivated activities designed to highlight ministerial visibility often took precedence over systemic change.

The ministers responsible for the Gender Affairs Division, as well as the gender-based technocrats within the Division at various stages (albeit for different reasons), noted that the rationale behind their activities was the need to be seen as "doing something for women." This modus operandi was also often explained as an example of political expediency, as programming is still the most effective means of bringing public recognition to the political minister in charge of the implementing unit. The endemic nature of clientelism in the larger political framework also exacerbates this need to "do for" women. As I began my interview with a political representative with responsibility for gender affairs, my two-hour wait was explained away by having to address the concerns and needs of the woman who had gone in before me:

> She came and said she wanted employment. She also said she wanted a house, somewhere to live because she lives with a relative, and then I said, "Okay." So, for the employment, I said, "What skill you have?" She said she had some. She's going to give me something with all the skills that she has. And she said with the housing, [interviewer has deleted

name] was working on something for her. I knew there was something more, and then she said, "But now, I don't have anything. I don't have groceries." So, right away, those are three different areas. (interview with then incumbent Minister for Gender Affairs April 2, 2002)

Here, the political representative was acknowledging this woman's three different immediate needs which the representative would try to immediately address. However, this preliminary exchange highlights the ways in which departments with responsibility for gender affairs can become an extension of welfare planning and provision of practical needs to the exclusion of strategic change (Molyneux 1985).

In an institutional context where political expediency trumps research and strategic planning, gender brokers are severely hampered in their ability to work systematically toward gender-just change. In response to whether a feasibility study had been conducted for one of the Division's projects, a former employee of the Division observed:

P.: Policy? It's not research driven; it's not from the bottom up. We stay in our office and the Minister may have some constituent in her head or someone who said, "You know what? I find you all should get involved in blah, blah, blah in Embacadere or somewhere, so you can do something for Embacadere, some women down there need training." And you go and you do something for Embacadere without any kind of data informing what you're supposed to be doing.

Rowley: And as a member of staff, did you feel "empowered" to speak back and ask, "Why?"

P.: It depends. After a while, you just did what you had to do. It was just easier, because you got all the licks about not being out there. She hated when we sat at our desk, she said, "You all are not out there, you all are not doing the work." And doing the work meant that you had to be in the field. Being in the field. (interview with member of the first staffing cohort March 18, 2002)

Conceptually, bureaucracies are not designed to bend to political will; *in practice*, yet, they do. The anxiety that is produced around securing subsequent terms in office makes gender-related focal points particularly vulnerable to co-optation, and activities are generated not around issues but through a sense of political expediency. Regional gender planning is, therefore, not driven by the demographics, need, or a vision of equity. Planning then becomes coterminous with the political intention of the Minister. Political visibility overwhelms policy direction, and these demands for directionless activity heighten during election years. The extent to which political expediency compromised a state-led vision for mainstreaming gender equity was best expressed by one former functionary of the Division who felt, during periods of high election campaigning, transformed into a worker in her Minister's campaign office rather than on behalf of a constituency of women.

Nonetheless, despite the articulated mandate, the historical pull of how women had been first institutionalized within the state had to be challenged if there could be any effective use of the in-house staffing capacity[18]:

> We were in a ministry, a social services ministry, and people wanted us to deliver. People wanted to find out what we were doing. So we couldn't be dabbling in policy, which is, you know . . . we were trying to put things in place in order to implement data. The Minister would say you all have to go out there and work. So, we worked like community development officers and piloted a lot of small projects. It's disheartening because you didn't have the funds to carry through and stuff like that . . . (interview with member of the first staffing cohort March 18, 2002)

In the first attempt at hiring staff who would be completely dedicated to the advancement of women's issues within the state machinery in 1994, the Division secured a range of skills areas, such as development studies, agriculture, economics, and accounting. The first director had also received post-graduate training in the area of Women and Development Studies. This combination of skills was poised to engage questions of policy, planning, and gender mainstreaming as their mandate issued by the Nairobi Forward Looking Strategies, the IDB, funders of the strengthening project, and the rhetoric of the state.

A recent posting for vacancies within the Division continued to move in the direction of redundancy in the call for applicants with training in the area of social work or sociology in which, for example, "Experience in the delivery of training programmes and/or community work is desirable. Training in gender issues is required. Familiarity with the concepts and methods of gender analysis and women's studies will be a definite asset" (*Daily Express* September 20, 2002, 16). In this posting, training in the skills of gender analysis appears tangential to the advertised vacancies, again revealing the lack of policy direction in the Division.

This oversight, while responding to an already existing institutional culture, is still inexcusable in light of a counter-discourse on the category "woman" generated by a strong feminist NGO lobby[19] and the fact that the Centre for Gender and Development Studies was established at UWI in 1993. Subsequent to its establishment, the department generated undergraduate minors in gender studies, gender and development, and a M.Phil./Ph.D. Programme in Gender and Development. Ten years later, despite an increasing pool of graduates with the capacity to do gender analysis, the posting that emanated from the Gender Affairs Division required that the successful candidate must be "trained" in the area of sociology and social work but possess merely "familiarity" in the areas of gender relevant expertise. Operationally, the importance of sharpening the Division's policy focus is integral to securing the Division's commitment to a gender-just society.

In 2002, the Division contracted The Centre for Gender and Development Studies, UWI, as the executing agency to spearhead the formulation of a National Gender Policy for Trinidad and Tobago with the hope that this process would institute more checks and balances in the area of gender planning. The responses to this have been mixed. One former employee of the Division captured this relationship between policy direction and political will most succinctly:

> Written policies are really a waste of time if the Minister has no intentions of carrying them out. To me, the more effective Ministries are the Ministries where you have strong Ministers and a strong P.S. (Permanent Secretary). It has nothing to do with the written policy. A lot of the policies that are implemented are unspoken and unwritten. It changes from year to year and from government to government and from personality to personality and that is how the Public Service operates. We don't follow any set policy. And when they come in, it depends on what are the topical issues, what are the main . . . I don't know, what are people clamoring about . . . those are the policies and what the Minister feels he or she is really hip with or bent on achieving. They could have written ten gender affairs policies. Nothing would have made a difference unless it was one that was fully sanctioned by the Minister and he or she was there for a length of time to see it implemented gradually. (interview with a member of the first staffing cohort March 20, 2002)

"They could have written ten gender affairs policies. Nothing would have a made a difference unless it was one that was fully sanctioned by the minister"—the level of astuteness here was prescient. Three years after the Gender Affairs Division's commission of said policy, the incumbent prime minister summarily dismissed it as having not been issued "by the Government and does not reflect Government policy. In fact, there are certain recommendations in the document to which the Government does not and will not subscribe. The Government is therefore requesting that the document which purports to be official Government policy be withdrawn from circulation."[20]

Narratives of family integrity have become a very successful mode of rejecting calls for reproductive and sexual rights. As the executing office, the Centre for Gender and Development Studies, UWI, responded to the policy's dismissal on behalf of the respective sector specialists. In an open letter, Acting Head Patricia Mohammed defended the policy by observing:

> Some issues in gender are undoubtedly controversial. Sexual and reproductive rights are not a euphemism for anything—it is exactly what it says, our human rights as they pertain to areas of our sexuality and reproduction. This is an area that has always concerned those who work in the area of gender. What Consultants in a gender policy are asked to do is to bring to the attention of the society at large, all the issues it must consider to ensure that the rights and freedoms of all of its citizens, in the

areas of work and employment, health and education, law and polity, and freedoms as ethnic groups, adherents to different faiths, or different sexes, as enshrined in the constitution of the land, are brought to the table for attention. If it does not do this, then we are remiss as experts in a field and in our mandate to act for and on behalf of the entire nation.[21]

A categorical dismissal of gender equity, when emanating from the seat of political power, is not just a dismissal of policy but of process and civil society. The political commandeering of gender-related policies and agendas reminds us that gender mainstreaming, as a form of state feminism, can at any moment be compelled to defer its mission to partisan political agendas. The term "Reproduction" appears in the revised gender policy only in the context of encouraging women's participation "in discussions on issues related to social reproduction and family friendly policies, and in the determination of expanded remedies for anti-discriminatory labour legislation" (2009, 21). The document is further introduced with a complete disavowal in bold that notes that:

The National Policy on Gender and Development does not provide measures dealing with or relating to the issues of termination of pregnancy, same-sex unions, homosexuality or sexual orientation. (ibid., 5)

While gender mainstreaming stands as one of the primary policy locations for a discussion and operationalization of what gender equity means in the region, this project continues to be hindered by institutional and political logics that undermine these goals. In the absence of popular or high-level support for a sustained focus on gender equity, the responsibility for maintaining some centrality on this front falls to the regional implementing and advocacy units.

Shifting analytical scales, as we have done, from the international, to the regional, institutional, and the personal brings the institutional contradictions and personal costs into greater relief. As we shift scales, the vision of international mainstreaming practices become increasingly mired in the politics, logics, and expediencies of the everyday. This contradictory quagmire is one where high level administrators point to the existence of bureaus as an indicator of their commitment to equity concerns while the stumbling block unleashed by these administrators can serve to render many of these bureaus ineffectual. These hindrances, I have argued, include the power of regional state-managers to veto decisions made within the bureaus, institutional locations that are hostile to gender equity, conceptual ambiguities that limit effective programming, and an ever-widening chasm between state-based feminist practitioners and feminist civil society. However, it is to the latter that I now turn my attention because it is in the shifting shape and scope of feminist civil society that I see the greatest possibilities for gender justice and equity in the Anglophone Caribbean.

4 Reproducing Citizenship
A 20/20 Vision of Women's Reproductive Rights and Equity

A substantive body of literature now makes very persuasive arguments for the importance of pursuing gender equity by means of redressing material inequities and the differentials that women experience within and from decision making and power wielding channels (Boserup 1965; Molyneux 1985; Barriteau 1995; Eisenstein 1996; Beall 1998; Elson 1998; Goetz 2007). This productive engagement notwithstanding, in this chapter I want to begin a discussion that I will sustain for the rest of *Feminist Advocacy*; namely, that gender equity in the Anglophone Caribbean must both re-engage and expand its attention to the materiality of the body through notions of embodied equity.

Rather than focusing on gender mainstreaming's limited successes regarding gender equity and women's access to material resources, I want to turn our attention to the analytical incapacities that emerge when faced with the shifting terrain of body politics in the Anglophone Caribbean. The beleaguered status of gender mainstreaming approaches within state machinery, once the explanation for many of mainstreaming's inadequacies, does not fully account for its incapacities regarding women's bodies, generally, and sexual minorities, more specifically.

I contend that these incapacities around the body are, of course, hindrances to the pursuit of equity and are better explained through an examination of three interconnected factors: the state's overwhelming fear of sexual autonomy and non-procreative sex (Alexander 1997); the increasing conservatism of the documents that now hold center stage in gender mainstreaming discussions; and finally, Caribbean state-managers' resistance to the idea that a campaign of social justice and rights stands as foundational to, rather than incommensurate with, ideas of development. In my subsequent chapters, I reach for a discussion of embodied equity by examining sexual harassment within the work place and rights for sexual minorities as issues that have become pivotal to feminist advocacy and equity in the contemporary Anglophone Caribbean.

In this chapter, I begin this engagement with the body by focusing on the issue of reproductive rights. Discussions of women's reproductive well-being tend to present advocacy strategies within the confines of the nation. I take a multi-tiered approach that explores the local, regional, and transnational nuances of abortion law reform and women's reproductive well-being.

At the local and regional level, I examine law reform in Barbados, a process which preceded the interventions of the population and reproductive rights platform of the International Conference on Population and Development (ICPD/Cairo) and the Beijing Platform for Action (Beijing). I move to a regional discussion in my examination of Guyana's abortion law reform, which itself looked regionally by building on the legislative advances that had occurred in Barbados. Thus, rather than examining advocacy strategies within the confines of a single nation, I explore the analytical nuances that a transnational perspective brings to local discussions of (reproductive) equity. Such a transnational perspective highlights the global and local intersections and the downward effects of policy positions enacted at the global level; it also allows us to incorporate a critique of the effects of colonization processes on contemporary debates. To this end, I examine the politics of navigating policy borders as feminist activists attempt to secure abortion law reform in Trinidad and St. Lucia in the context of a reproductive rights landscape that is framed by the rhetoric and contradictory imperatives of Cairo, Beijing, and the Millennium Development Goals (MDGs).

EMBODIED EQUITY: PIECING THE BODY APART

What are the stakes of what I'm referring to as embodied equity? My investments in centering the body as a critical analytic in this project's discussion of gender equity were reinforced after a very rewarding evening of intellectual sparring with my graduate students when the body reappeared. It was a specific body, one that resisted the language of "haunting" and "ghostliness" that had fed the palpable excitement that I had in earlier discussions with my students. This body resisted deconstruction and called for its re/membering in light of the terror of dismemberment that it had experienced. This body came to me over the radio air waves as the announcer recounted the systematic rape of women who had opposed Zimbabwean President Robert Mugabe's Zimbabwe African National Union-Patriotic Front (ZANU-PF) party.[1] The written account of these human rights violations, "Electing to Rape: Sexual Terror in Mugabe's Zimbabwe," points to the systematic rape of women who belonged to the opposing political party, Movement for Democratic Change (MDC). Based on interviews with seventy-two survivors, the report conservatively estimates that in the run up to the 2008 national elections in Zimbabwe 380 rapes were committed by 241 perpetrators across Zimbabwe's ten provinces, with each woman being raped on average five times (2009, 17).

The sensations that forced me to feel my own corporeality have stayed with me and have prompted me to rethink how the body matters in debates about development and gender equity. The idea of embodied equity which guides the next three chapters of *Feminist Advocacy* aims to get at a very particular set of concerns. Namely, what is required so that the physical body is kept safe from harm and violence? What are the discursive frames that render violence against the body as reasonable and necessary? How do we intervene within these frameworks so that differently situated bodies are able to rightfully assert claims to the nation?

To answer these questions, we must temper any inclination to see the body as a mere fiction, but we also need to do so in a way that remains mindful of the discourses that are mapped on the body to legitimize various forms of vulnerability and harm. I am thinking of embodiment, therefore, as a way of capturing the intimate and divergent interplay of the physical body as both text and writing instrument. Embodiment gestures toward the ways that lived bodies are called forth and made to signify differentially within social, cultural, historical, economic, and political frameworks— they are written upon as text. However, embodiment also points to the ways in which these individual bodies engage and inscribe the social collective— writing their subjectivity on the social order. My discussion of embodiment draws on two very disparate discussions, anthropologist Mary Douglas's *Natural Symbols: Explorations in Cosmology* (2003) and feminist philosopher Rosi Braidotti's *Nomadic Subjects* (1994).

Douglas's early work highlights the body as a symbolic expression of the social order. In Douglas's work, the body has no natural bearing or meaning. Rather, it takes its value from the social categories available to it. She writes:

> The social body constrains the way that the physical body is perceived. The physical experience of the body, always modified by the social categories through which it is known, sustains a particular view of society. There is a continual exchange of meanings between the two kinds of bodily experience so that each reinforces the categories of the other. As a result of this interaction, the body itself is a highly restricted medium of expression. The form it adopts in movement and repose expresses social pressures in manifold ways. The care that is given to it, in grooming, feeding and therapy, the theories about its needs by way of sleep and exercise, about the stages it should go through, the pains it can stand, its span of life, all the cultural categories in which society is perceived, must correlate closely with the categories in which society is seen in so far as these also draw on the same culturally processed idea of the body. (72)

Douglas's discussion of the body lends an important insight into my discussion of embodiment. The body emerges as text that is circumscribed by social meaning, its physical processes rendered meaningful through narrative and ritual. The way that the body comes to matter in Douglas's work does, however, provide some limitation in that it does not quite capture what I refer to as the body's "divergent interplay as text and writing instrument."

For example, Douglas neither captures the ways in which different bodies are differently inscribed (it does not account for how social meanings vary culturally and within sub-cultures), nor does it address the political and moral hierarchies that attend these social meanings. Moreover, while the body is a restricted medium of expression, it nonetheless *remains* a medium of expression—a writing instrument that also infuses meaning into the social body thereby producing a degree of dynamism to the social collective.[2]

Braidotti's discussion, however, points to embodiment in which the body emerges as "the material, concrete effect, that is to say, as one of the terms in a process of which knowledge and power are the main poles" (57). Embodiment within Braidotti's framework points to practices that discipline, normalize, and control the body but also what she refers to as a messy "cartography" of inter- and intra-differentiated experiences of the body (158). While I draw on Douglas for an understanding of an embodied self that is attentive to and participates in a variety of rituals for grounding and sense making, I find Braidotti's careful location of inter/intra group differences in the experience of the body invaluable. Similarly, her acknowledgment of how our sense-making mechanisms occurs through practices of control of the body is critical to my use of the term "embodied" (Braidotti, 57).

The physical body's experience of structure, symbols, rituals, power, and messy relationality is, therefore, what my shorthand use of "embodied" gestures toward. Additionally, in *Feminist Advocacy* embodied equity names a mode of gender equity that aims to interrogate and redress the specific ways in which differently situated and marked bodies are subject to narratives and rituals, called forth and rendered hyper(in)visible through a logic of marginality. By coupling "equity" with "embodied," I aim to "re/member" the body, not in a way that returns us to a Cartesian rationality or unity, but rather to take notice of differently situated bodies and differently located sites of pain and pleasure so that we can better intervene in the ways these bodies have been pieced apart by social practices of excision.

MAINSTREAMING GENDER/REPRODUCING THE MAINSTREAM: IMPLICATIONS FOR THE NEW MILLENNIUM

In this chapter, I focus specifically on the reproductive politics attached to women's bodies in the Caribbean. I explore the constraints of reproductive rights advocacy in the context of the increasing conservatism of mainstreaming documents and international policies that speak to reproductive health and well being. The effects of recent and rising ideological conservatism within the transnational arena remain the core focus of this chapter. This conservatism is felt more acutely as we develop greater analytical insight into the ways in which the shifting terrain of gender justice is an embodied one. Advocates who pursue this form of embodied equity will find that there is little to support their efforts in any of the primary policy documents and approaches that presently undergird gender mainstreaming efforts. I say this with some qualification because where abortion law reform has occurred in the Caribbean, for example in Barbados and Guyana, it occurred prior to the adoption of important reproductive rights platforms such as the 1994 ICPD/Cairo and the 1995 Beijing Platform for Action (Beijing). As such, abortion law reformers might argue that these documents have always been irrelevant to the type of change that is possible in the region.[3] However, the contemporary period of mainstreaming, governed as it is by these documents as well as the MDGs, ironically presents a political landscape that is increasingly recalcitrant and unresponsive to feminist demands for change as is proved by the failure to obtain abortion law reform in Trinidad and Tobago and in St. Lucia.

Where broad abortion law reform has occurred in the region, we find that women's bodies have been mainstreamed into a public health, macro-economic framework that is instrumental in nature, dependent on widespread debate and consultations, and predicated on the work of transformative state agents and allies vested with cultural and political capital recognizable to the mainstream political and social order. These features are particularly important on two grounds. First, they home in on the idiosyncrasy of small states where individual credibility matters significantly to the nature of transformation (or resistance) possible. Second, they point to the ways in which intelligibility to the mainstream produces a peculiar sleight of hand in which transformation maintains a veneer of not disturbing the status quo.

When equity debates turn to women's reproductive rights, gender mainstreaming equality policies and state policies increasingly look like mirror images of each other, leaving progressive advocates even more hard pressed to work on behalf of a feminist lobby. Gender mainstreaming as a form of state based equity finds itself in a careful and, at times, too tentative negotiation with the state's fear of women's sexual autonomy, thereby undermining the revolutionary possibilities of gender mainstreaming from within state machinery (Alexander 1997, 64).

It is no accident that women's reproductive rights and the sexual autonomy of women and sexual minorities, matters that are profoundly connected to an equity platform, find little support in contemporary gender mainstreaming frameworks. This results largely because gender mainstreaming remains tethered to the primary impulses of state-machinery to filter sex and sexuality through a lens of procreation, population control, and by extension, the management of domesticity. Consequently, pleasure, bodily autonomy, and individual desire are not only reviled but actively resisted for having no productive purposes. This persists despite the fact that feminists have long pointed to the political economy of a wide range of sex practices from that which occurs in marriage to the informal economy of sex work. The excision of rights language and the marginalization of women's sexual autonomy and sexual minorities are all symptomatic of the conservative ideological bent resultant from the various sovereign and religious influences on UN policy formulation.

Legal theorists Rebecca Cook and Bernard Dickens, in their article "Human Rights Dynamics of Abortion Law Reform" (2003), designate reproductive rights as critical to women's comprehensive participation and ownership of their citizenship and suggest that the tide is turning in legislative understandings of this connection. Their women's/human rights framework includes the decriminalization of abortion, the provision of public health services to minimize maternal health mortality, and the acknowledgment of women's reproductive autonomy as linchpins in the processes of social justice, citizenship, and democracy (6–7). The authors also acknowledge that a human rights framework requires feasible realization possibilities alongside of the appropriate legal instruments. As such, what is notable about this human rights framework is the authors' insistence that the levels of accessibility that women have to pre- and post-natal health care matter equally as does the need for the legislative and medical infrastructure so that women are not coerced into positions that would be detrimental to their own well-being. They note:

> Legislatures and judiciaries respectful of women's views, including those that hear women's voices from within their own memberships, are progressively molding legislation and its interpretation sympathetically to women's interests in health and in observance of human rights. As women become equal citizens with men in their societies, it is anticipated that abortion concerns will evolve from placement within criminal or penal codes, to placement within health or public health legislation, and eventually to submergence within laws serving goals of human rights, social justice, and the individual dignity of control over one's own body. (ibid., 7)

Cook and Dickens's singular deployment of the category "woman" may be reminiscent of early approaches that analyzed women as a homogeneous social category. This is not an unreasonable deployment in as much as their use of the category "woman" rightly allows us to think in terms of how women as a reproductive unit may be affected by an expansion of reproductive rights legislation.

Cook and Dickens's vision for the expansion of and interconnections between reproductive equity and human rights is one that I embrace. Still, it is also a vision that I think demands some degree of interrogation even as we work toward its realization. Despite the fervor of the debates and the resulting camps that ensue, the irony of decriminalization legislation is that it could be seen as neutral in its impact as it neither does harm nor does it compel anyone to act in one way or the other. In other words, decriminalization in no way causes harm to women who choose not to have an abortion. Further, decriminalization in no way compromises the citizenship of women who do not see the terrain of their equality as connected to reproductive rights. What decriminalization legislation does is make it possible for women who wish to terminate their pregnancy to do so. Consequently, the citizenship of those who never invoke the legislation is not compromised, and those who are enabled by it can continue to participate in the national project as law-abiding citizens rather than be excised as criminals. Abortion law reform, mired as it is in gendered politics and culture, is never this simple.

There is a tendency within the literature to suggest, with some degree of repetition, that the absence of legislative shifts toward decriminalization is the result of judiciaries, parliaments, and other masculine-ruling elites' wielding culture. To be dismissive of such power blocs is foolhardy. However, we should remain equally cognizant of the ways in which women's participation and at times complicity with culture can be used by the status quo (of which the ruling-male elite is a subset) to undermine gender equity. Women co-write culture, admittedly with different authorial powers. Still, such influence, as will become apparent later in this chapter, is highly contextual. To dismiss these interactions positions women as victims of culture, a thing that can only be done to them. Ignoring women's complicity with the writing of culture functions as the corollary of feminist philosopher Uma Narayan's discussion of how women who critique their culture are labeled as "inauthentic." The label of inauthenticity serves, Narayan argues, to position them as alien and their criticism, at best irrelevant, at worst acts of betrayal (1997). Regardless of which side of the equation, whether women's complicity is ignored or their criticism made suspect, the intended effect is the same: a preservation of the status quo.

Working on behalf of reproductive equity in its entirety contradictorily means sometimes opposing those who form part of the constituency to whom this right will be applied. In other words, this is a struggle in which the acquisition of women's rights may oppose what a number of women think to be right. However, Cook and Dickens's call to shift reproductive equity from a discussion of penal codes to that of public health is potentially one of the most effective ways of speaking across these rigid pro-life/choice divide. The move from the realm of law to public health does not totally eliminate the ideological fervor that attends these debates. It would be ideal to simply name abortion a medical procedure; yet, this would require ignoring the many cultural markings that rest upon how we see abortions. The shift from the legal to the public health realm, though, does facilitate a more neutral discussion about women's right to make health-related decisions about their reproductive physiology as they do about other anatomical aspects of their lives.

My earlier distinction between equality and equity becomes germane to the discussion at this point. Men and women achieve reproductive citizenship through vastly different channels and with significantly different sociopolitical attention on their actions and choices. Consequently, women's needs of the health care system are drastically different. Embodied equity singularly dictates that we attend to the required transformations that allow women to function—and be seen as functioning—as rational actors in relation to their bodies. The legal and medical transformations that are required for women to be seen as rational functioning agents hold the radical potential of envisioning women's control over their bodies as a basic human need.

Much of our understanding of reproductive advocacy is indebted to the transnational dialogues which occurred among national governments and women's NGOs in the mid-1990s. The 1994 ICPD in Cairo and the 1995 Beijing Platform for Action have been critical to both the expansion and consolidation of our understanding of reproductive equity. Coming on the heels of the 1993 Vienna Human Rights Conference, these two world conferences, more than any other transnational activity, have advanced our conceptual understanding of reproductive health. Together, the policy outcomes of Cairo and Beijing, by defining the terms of reproductive equity, by naming and establishing a clear recognition of the constituencies to whom such conceptual parameters matter (e.g., adolescent girls), and by shifting the reproductive debates from a demographically informed discussion of population development to questions of reproductive well-being, have centralized the importance of crafting legal and medical infrastructural synergy toward an alignment of reproductive equity with women's rights as human rights.

Prior to the ICPD and Beijing, the CEDAW in 1979 stood as one of the few UN documents to speak to the importance of women's health and reproductive equity. Article 12 encouraged state parties to "take all appropriate measures to eliminate discrimination against women in the field of health care in order to ensure, on a basis of equality of men and women, access to health care services, including those related to family planning." It is useful, therefore, to read the positions emanating from Cairo and Beijing in tandem with each other to fully grasp the extent to which these two conferences advanced our understanding of the importance of reproductive equity.

Cairo signaled a number of turning points for women on reproductive rights. Political scientist Paige Whaley-Eager notes that prior to Cairo the very use of "reproductive rights" within UN policy documents was virtually absent (2004, 147). Prioritizing women's personhood, Article 7.2 of the Cairo Program of Action highlighted reproductive health as "a state of complete physical, mental and social well-being and not merely the absence of disease or infirmity, in all matters relating to the reproductive system and to its functions and processes." This definition, repeated in Article 96 of the Beijing Platform, stated that "Reproductive health therefore implies that people are able to have a satisfying and safe sex life and that they have the capability to reproduce and the freedom to decide, if, when and how often to do so."

Whaley-Eager points out that the Cairo conceptualization of reproductive health was one that pushed the reproductive discourse from an instrumental management of women's reproduction within national/global population control regimes to a recognition of women's agency and autonomy over their bodies (2004, 146).

By expanding the correlation between sex and procreation to consider factors such as the importance of a "satisfying sex life," Cairo and Beijing's conceptualization of reproductive health has also opened an equally as important opportunity for us to consider matters of embodied pleasure and women's right to experience pleasure. Arguably, this consideration of a "satisfying sex life" can be seen, too, as an implied indictment of practices which would impair enjoyment of sex. Expectedly, the documents emanating from these conferences were each mired in their own politically fraught moments, characterized by chasms, dissent, and the need for skillful diplomacy. The International Women's Health Coalition, in a piece written by their senior administrators, reminds us that Beijing began with 35 percent of the platform for women's rights bracketed in contestation (Dunlop et al. 1996). Such opposition to women's rights reflected an offense mounted by what Stean and Ahmadi referred to as the "pragmatic and expedient alliances" of conservative and religious fundamentalist forces from the West and the Middle East (2005, 229). Skating dangerously close to reversing some of the agreements made one year earlier at Cairo, diplomats and activists acting on behalf of Beijing's core commitments succeeded in reaching what Joan Dunlop et al. have lauded as some of the strongest international language on women's rights as human rights and the importance of women's autonomous control of fertility and their social and political empowerment (1996, 158).

Mention of abortion, however, was limited to a narrow but disproportionately controversial aspect of reproductive health. Cairo and Beijing encouraged governments to improve women's access to family planning services and information in order to minimize their recourse to termination procedures (paragraph 106 [k]), a matter which was first iterated within the ICPD Program of Action (Article 7.24). Despite the ICPD's notable progress in advancing a rights discourse around women's reproductive autonomy, it is worryingly conciliatory in its language of best practices. Throughout the document, powerful assertions made regarding the need for legal reform, counseling, and education are immediately tempered with the Program's recurring mantra that all recommendations remain subject to the "sovereign right of each country consistent with national laws and development priorities, religious and ethical values and cultural backgrounds of its people, and in conformity with universally recognized international human rights" (ICPD, Preamble).

The Health Expert Group Meeting, which met in 1998 in Tunis, Tunisia, offered possibly one of the most transparent statements on the value of decriminalizing abortion to be found within the UN's aegis. The Expert Group met with the stated purpose of addressing the Beijing Platform's mandate to mainstream a gender perspective in every sector, including health. The delegates present called upon governments and members of the international community to "address the reality and consequences of unsafe abortion by revising and modifying laws and policies which perpetuate damage to women's health, loss of life and violation of gender equality in health care" (Article 57). Yet, this, too, was followed by the recurring mantra of adhering to the primacy of national statutes and law (Article 58).

As an amalgamation of sovereign states, the UN is expectedly respectful of state sovereignty. However, this conciliatory tone truly reflects a conversation between and among patriarchies about women's reproductive well-being, the strength of religious fundamentalism and ideological conservatism within UN processes, and the heteropatriarchy of masculinized nation-states. These conversations prompt us to question the extent to which gender mainstreaming possesses the capacity to facilitate national advances in the area of women's reproductive rights. They also mark an unusual concession for the primacy not only of law, but also of culture, tradition, and religion rendered in ways that suggest that these elements inhabit a supra-gender location rather than, as we know them to be, deeply gendered and discriminatory locations—more so because of their aura of neutrality. Despite the advances of Cairo and Beijing on reproductive rights, a resurgence of paternalism between transnational and national conversations leaves women with very little access to the resource of moral suasion to facilitate the very mainstreaming objectives that are desired within all sectors of society as advocated by Beijing. Such concessions can facilitate an oppressive majoritarianism, which when rendered as democracy, suppresses the rights of minority groups under the guise of justice.

Though qualitatively different, the UN's tentative reproductive equity policy position is, as with hostile, restrictive national legislation, distinctly at odds with the kinds of daily decisions that the world's women make about their reproductive lives. In some cases, women's daily decisions are made as a free choice among a range of options; in others, *de facto* actions resulting from an absence of options. These two scenarios, of course, provide women with distinctly different outcomes when intersected with class, geopolitics, access to information, and an ability to act on such information in intimate contexts.

Within the developing world, maternal mortality is a telling and preventable indicator of this disjuncture between restrictive legislative decisions on reproductive health and equity and women's daily lived "choices." The World Health Organization's (WHO) statistics on maternal mortality are a stark reminder that women are the most vulnerable casualty of legal and medical failures on reproductive health and equity.[4] According to the WHO, approximately 1,500 women die from pregnancy- or childbirth-related complications on a daily basis. Further, in 2005, approximately 536,000 women died from childbearing-related complications, a burden that disproportionately affects 99 percent of women in the "developing" world.

Certainly, a genuine commitment to reducing maternal mortality as advocated by Cairo and the MDGs would dictate that we collectively confront the fact that approximately 13 percent of these deaths occur from complications of unsafe abortions (WHO). What this means is that in 2005 approximately 69,680 women around the world died from complications resulting from an unsafe abortion.[5] This figure is potentially an underestimation if we bear in mind the stigma and legal implications for many of the world's women if they actually did report an attempted abortion. Researcher Susheela Singh, in a study of thirteen developing countries,[6] looked at hospital admissions resulting from unsafe abortions and estimated that approximately 5 million women are admitted to hospitals for complications resulting from unsafe induced abortions. Criminalizing women for the reproductive choices that they make initiates an attending network of dysfunctions, ranging from the foreclosure of discussions about women's reproductive lives to their stigmatization and death. Expectedly, Singh calls for increasing decriminalization but acknowledges that decriminalization requires an equally comprehensive set of supports inclusive of education on the implications of unsafe abortion, an awareness of the related economic costs, and a numerical awareness of the numbers of women who are adversely affected by the effects of criminalization (2006, 1891).

THE MILLENNIUM DEVELOPMENT GOALS AND THEIR REPRODUCTIVE CONTRADICTIONS

I now turn my attention to the UN's MDGs as another significant obstacle wherein gender mainstreaming frameworks are constrained by existing policy documents to address embodied equity claims. The "new" developmental mode of poverty eradication is best reflected in the much touted MDGs.[7] On September 6, 2000, 191 UN member states met in New York and on September 8, 2000, 189 of these states agreed to adopt the Millennium Declaration. Legal theorist Philip Alston argues that the world's leaders have embraced the vision of the MDGs because, unlike their predecessors, the MDGs possess four important strengths: (1) they avoid an extensive shopping list of desirable goals by prioritizing those that are most pressing; (2) they are numerically measurable and increase opportunities for both measurement of achievement and accountability; (3) with a culminating year of 2020, they are time bound, and, finally; (4) they have received unprecedented institutional support toward their achievement (2005, 756). Alston's assessment offers some explanation for the rapidity with which many of the world's governments have incorporated the MDGs' objectives into their own national management plans. It is disturbing that the present perspective most vivid in the international imagination regarding development, namely the MDGs, is exceedingly weak on women's reproductive options.

The widespread embrace of the MDGs within the national strategic plans of many of the world's governments leads me to think of the MDGs as a form of "globalized developmentalism." I use this term to refer not only to the expansive embrace of the MDGs within the national strategic plans of many of the world's governments but also to the practices and ideologies that shape the *marketing* of prescriptive notions of "development" by what Alston refers to as "extensive institutional apparatus that has been set up to promote them." I use "globalized developmentalism"[8] to reflect a Fordist cooker-cutter model of producing development through the pursuit of eight MDGs. The emphasis here is on the "marketing" of "development" as a branded product that, despite the difference in parameters, resonates in very similar ways within and by populist organizations, such as one.org, as it does in more formal transnational organizations, such as the UN and the Bretton Woods institutions.

This is especially true in the Anglophone Caribbean where the MDGs have captured the imagination of government officials. This is not, however, out of a commitment to equity, but as a resuscitated modernization paradigm, for these goals channel familiar memories and desires to be a recognized actor in the "modern" world.[9] This desire for modernity is not inherently problematic. It becomes a troubling fantasy, though, when it is interpreted in ways that privilege market forces without a simultaneous (con)frontal engagement with the challenges of socio-political equity and an engagement with those who have been labeled as abject within our societies.

There is a recurring absence of attention to women's reproductive health within the region's list of MDG priorities. There is, and appropriately so, an emphasis on poverty reduction, HIV/AIDS, and improving data collection capacity (Caribbean Development Bank 2005). The danger of not employing an intersectional analysis to the MDGs is evident by the ways in which women's reproductive well-being recedes even along those axes that would benefit from a discussion of women's reproductive well-being (e.g., poverty reduction among female heads of households). Jamaica breaks with this regional pattern by incorporating teenage pregnancy and consequent dropout rate for girls as one of its MDG priorities, a need that results from the tendency of some schools to refuse re-entry to adolescents who have become pregnant. Jamaica has also signalled an intent to address maternal health data collection as one of their priorities (ibid.).

In the best of all worlds, the erasure of women's reproductive health from the Caribbean's operationalization of the MDGs would suggest that all was well for women and their reproductive well-being. Not only is this not the case, but also the MDGs' emphasis on numerical markers produces a development culture in which prioritizations and hierarchies trump interconnections between and among crises. This has re-ordered developmental priorities in ways that make it possible and legitimate for women's reproductive well-being, if recognized, to be recognized primarily in terms of crises and emergency management.[10] For example, rather than recognize the interconnections between poverty or HIV/AIDS and women's reproductive rights as a pillar of social justice, we are repeatedly asked to prioritize one above the other.

Arguably, if we operationalize the MDGs in particular geopolitical contexts, what Alston identifies as one of their strengths, namely their measurability, may emerge, too, as a persistent weakness. One of the potential failures is precisely the need for standardized measures and indicators of success. Rather than producing a tiered measurement, which assesses achievement based on where states are presently, the MDGs themselves offer indicators, which for many states (although not enough) have already been achieved (e.g., 50 percent reduction of individuals who live on less than one United States dollar/day, universal primary education, and reduction of gender disparity through educational access).

The Caribbean, for example, remains a remarkably heterogeneous economic space. In 2000, the per capita GDP ranged from US$1,467 to US$17,012 (Bourne 2003).[11] However, these figures are deceptive, as they tell us very little about the distributive element of a country's resources. Trinidad and Tobago has a per capita income of US$ 18,864 (2008), placing it significantly above the MDG's goal of reducing the number of people who earn one United States dollar/day. Yet, only 26 percent of the country receives a daily supply of pipe-borne water, with an estimated need for $1.2 billon over three years to increase the coverage to a mere 36 percent (*Sunday Express* August 7, 2005).

I highlight this disparity to identify the limitations of indicators, namely, their incapacity to adequately reflect the complexities of distribution and access. As such, the region's embrace of the MDGs cannot merely be a return to a Basic Needs approach of poverty eradication, little more than a resuscitated modernization paradigm within which people function as part of an efficiency framework, i.e., as the means to the best set of outcomes. The MDGs, in the midst of addressing the exigencies of life, must be tempered to acknowledge that as persons we exist in need of, but also beyond, these immediate goods. Caribbean feminist and former General Secretary of DAWN, Peggy Antrobus, expressed such sentiments:

> You don't want to say that all people really need is food and shelter. It's really beyond basic needs ... I think for women you can't separate the rights and dignity issue, in the same way you can't separate

women's reproductive rights from women's health. You can't separate freedom from want and freedom from fear if we think of sexual harassment when a woman goes to get a job, for example. This is why I think the vision for the women's movement is much more holistic because we know that women's lives cannot be compartmentalized. We can't talk about these things without talking about other things. (Rowley 2007, 76)

This chapter, therefore, is an exploration of what these "other things"—as they extend beyond basic needs—might be in the context of women's reproductive health and equity.

This chapter directly foregrounds an ethics of development to centrally address the ways that disadvantaged groups and silenced constituencies are able to lay hold of the collective resources of the state (e.g., fiscal, legislative, political, and social). An ethics of development should, then, pursue ways of evaluating the impact that these resources make on the material well-being of disparate individuals *and* aim to enhance their capacity to pursue a range of life possibilities. To this end, does the present landscape suggest to us that, without change, women in this futuristic model will be participating agents in debates about reproductive health and sexual well-being?[12] Are these in themselves desirable goals? And, if no, what organizational and ideological parameters might we adopt and implement to make the question of equity an achievable goal for women via the MDGs?

ENVISIONING DEVELOPMENT: WHAT DOES THE YEAR 2020 HOLD FOR CARIBBEAN WOMEN?

That Caribbean state-managers, in their embrace of the MDGs, are not compelled to think in complex ways about women's embodied equity in relation to the MDGs is to be expected in light of the fact that the MDGs do not demand sophisticated understanding of how the lived body is implicated in development processes. Women's equality is centrally addressed in Goal Three, which articulates the need to "promote gender equality and empower women by eliminating gender disparity in primary and secondary education preferably by 2005 and at all levels by 2015." Naila Kabeer reminds us that this is an "intrinsic rather than instrumental goal, explicitly valued as an end in itself rather than as an instrument for achieving other goals" (2005, 13). This is an important distinction. Kabeer is not suggesting that improving women's access to education will not facilitate instrumental gains. For example, increased participation in the labor market or reduced fertility levels remains beneficial to individual women even if these benefits did not translate into gains for the broader social collective.

This is a position that is well argued by Pulitzer Prize-winning development economist Amartya Sen in his important work, *Development as Freedom*. Sen's work is an exploration of development's ability to offer many of the world's people a more "unfettered" life. As a mediated position between the critique of developmentalism and a total embrace of development as "growth/good," Sen approaches development paradigms with the understanding that neither are they an axiomatic good, nor will the most marginalized secure benefits from economic growth. Sen writes:

> Individual freedom is quintessentially a social product, and there is a two-way relation between (1) social arrangements to expand individual freedoms and (2) the use of individual freedoms not only to improve the respective lives but also to make social arrangements more appropriate and effective. (2000, 31)

In Sen's work, development approaches are at their most productive moment when citizens' freedoms are treated as the ends and means of development. This is a carefully calibrated balance in which social justice facilitates and enables people's making use of their individual freedoms (31), economic and social platforms are symbiotic pursuits (47), and one ought not to preclude the other (7). Oddly, Sen discusses the expansion of women's agency in terms of its enhancement of other "classic variables," such as education, reducing child mortality, intra-household equity, and reduction in fertility rates when he notes that "these matters have general developmental interest that goes well beyond the pursuit specifically of female well-being, though—as we have seen—female well-being is directly involved as well and has a crucial intermediating role in enhancing these general achievements" (203).

Sen's observations are consistent with his overall premise that the pursuit of all freedoms reinforces and enhances the realization of other freedoms. However, the distinction between intrinsic and instrumental goals is a distinction that prompts me to think of the ways in which intrinsic goals can be put to instrumentalist purposes and the implications of such for women. I want to explore this possibility in the context of the MDGs and women's reproductive well-being. Without a guiding ethic that maintains a sustained focus on the question of women's value, the dictates of market and political pragmatism often prioritize instrumental gains above intrinsic ones. The MDGs, without a substantive philosophy that addresses the intrinsic value of those on whose behalf they aim to work, are particularly vulnerable to such manipulations.

I want to suggest that such vulnerabilities are present for women for three primary reasons. First, they draw on important yet inadequate and narrowly defined indicators of gender equality and are, therefore, unable to account for the nuances of gender asymmetries. As Alston notes, the measurability of the MDG indicators is a decided strength that led to their overwhelming embrace, but these indicators do not adequately address individual outcomes or the ideological factors that can hinder full access for people based on their differentiated position within society, matters which, of course, lie at the heart of gender inequity. Second, when the MDGs do address women, they prioritize a maternal identity. This is not a criticism of a "maternal identity," although I do think that we need to challenge the MDGs to maintain a consistent gender analysis throughout the operationalization process. I am, though, concerned that the MDGs' attention to maternal mortality remains disconnected from the factor that is significantly responsible for maternal mortality worldwide, that of complications from unsafe abortion. Third, I think that women are undone by the lack of an ontological and embodied perspective. The MDGs, as a result of their instrumentalist formulation, do not concretely address the importance of subjecthood, human value, or the ways in which this value is lived through the body. While this is implied in the MDGs' commitment to improving the livelihood of the world's marginalized, I remain skeptical of the sustainability of instrumental approaches that do not adequately incorporate a recognition of the individual's intrinsic value.

Additionally, the MDGs offer reproductive activists little to stand on because they allow for intra-organizational amnesia by virtue of not building on the feminist antecedents inherent in documents and platforms such as CEDAW, GEM, and Beijing or Cairo. Despite these documents' constrained perspectives on reproductive rights, they at least offered a feminist vision of equity. By circumnavigating the questions of value that feed these antecedents, the MDGs have been sanitized of any feminist intent. Feminist possibilities of change have been replaced with a framework of gender equality that resonates all too easily with the achievement of basic needs, rather than, or in dialogue with, locally specific conditions of empowerment. Ceri Hayes, for example, in an analysis of the ways in which the MDGs might be enhanced by CEDAW, observed:

> Looking at the MDGs through a CEDAW lens adds another dimension to these arguments. The Convention rests on the conviction that all women have human *rights*, not just *needs*. Seen in this light, the ideals of equality and non-discrimination are, in fact, important ends in themselves, not simply a means of delivering the MDGs in a cost effective way (2005, 70).

A strong feminist critique of the MDGs' capacity to envision significant transformative realities for the world's women by the year 2020 is not only necessary but also dire. The limitations of the MDGs are most glaringly evident in their perplexing exclusion of indicators for reproductive health and well-being, an exclusion all the more perplexing precisely because women's reproductive health and well-being not only are inherently important goals in themselves but are critical to the achievement of other stated MDGs (Antrobus 2005, 95). The MDGs' approach to women's reproductive health and well-being (re)produces a procreative identity as *the* normative understanding of women's reproductive well-being to the exclusion of the ICPD's antecedent of rights, decision making, and satisfaction that craft a more holistic subjectivity for women.

Millennium Development Goals Three, Four, and Five—educational access and reduction of child mortality and of maternal mortality—in tandem with each other signal a maternal triumvirate. That women's fertility rates are reduced when they receive access to education has now become the pat line of argumentation used by policymakers when arguing the value of educating women. This perspective, as we have already discussed, is at the heart of MDG Three. However, the maternal element is further crystallized in MDGs Four and Five: the latter drawing attention to the need to reduce maternal mortality by 75 percent; and the former, the need to reduce child mortality for children below the age of five by 66 percent.

It is not that these are not meritorious goals. They are, though, certainly not new ones.[13] In the context of the Anglophone Caribbean, the additional and peculiar difficulties of data collection stymied these goals. Among the peculiarities is the fact that maternal mortality is calculated as maternal deaths for every 100,000 live births, which then produces statistically low results for maternal mortality rates in the region since, for many countries in the Anglophone Caribbean, live births per annum often does not exceed 10,000.[14] Similarly, because of the disparate ways in which data is collected in respective territories, standardization has to occur with some degree of urgency before collective and comparative discussions are possible. These combined factors make maternal mortality a precarious indicator of women's equality in the Anglophone Caribbean.

Further, when a woman's identity as *mother*—as reflected tangentially in MDG Three and centrally in Goals Four and Five—is the *only* parameter used to reflect women's possibilities, then both parameter and possibility become dreadfully narrow in their contextual operationalization. Take, for example, the sentiment pertaining to maternal mortality that emanates from a report on the MGDs submitted to the Caribbean Development Bank:

> The argument for saving women's lives is two-fold. Not only is it important in itself, but also because of the negative consequences of a woman's absence on the well-being of households in general, and children in particular. Research has shown that households with women tend to do better than households without, since women are more likely than men to utilise greater proportions of available resources for the benefit of the household and especially for children (2005, 29).

While the report points to the inherent good of saving women's lives, it reflects also one of the detrimental consequences that can result from crafting instrumental arguments about women's reproductive possibilities: the problem of dismembered citizenship.[15] Cultural norms often mandate women's disproportionate allocation of resources (e.g., food, time, and social-capital) to the household at great sacrifice or loss to themselves (e.g., cultural norms where men/male children eat first and most). Anything less than a holistic analysis aimed at securing women's well-being tends to lose sight of these important distinctions.

Admittedly, a strong transnational feminist critique has resulted in the MDGs' being operationalized within a more robust framework of reproductive health, which recasts links with ICPD and CEDAW. Authors such as Noeleen Heyzer, Peggy Antrobus, and Naila Kabeer (2005) have systemically assessed the areas of health care, education, and gender cultures, interpreting each goal through the lens of reproductive health.[16] The sentiment inherent in this feminist critique received a strong endorsement from then UN Secretary General Kofi Annan who, two years after the adoption of the MDGs in 2000, observed:

> The Millennium Development Goals, particularly the eradication of poverty and hunger, cannot be achieved if questions of population and reproductive health are not squarely addressed. And that means efforts to promote women's rights, and greater investment in education and health, including reproductive health and family planning.[17]

The importance of a multi-sector critique with a commensurate regard for the interconnections that exist between justice and economics and between human progress and value lies at the heart of Annan's observation. This intersectional critique is one which presses against the measurement-driven impetus of the MDGs.

Demographer David Yeboah notes that as Family Planning Association (FPA) organizations in the Caribbean shift from state-driven population control models to services, their funding has been heavily subsidized by external sources, such as International Planned Parenthood Federation, United States Aid for International Development, and CIDA (2005, 1). This dependency produces a degree of vulnerability for the region's FPAs, which then become subject to the flows and ebbs of the political economy of the North. In the midst of this momentum regarding reproductive health, an interesting alliance emerged that has particular resonance for the Anglophone Caribbean, namely, the couching of reproductive health in the context of HIV/AIDS.[18] With an awareness of the increasing prevalence of HIV/AIDS in the region, Yeboah observes that the reduction in international funding has crippled the region's capacity to fight

HIV/AIDS (ibid.). What the MDGs now make possible, through the coupling of reproductive rhetoric with HIV/AIDS, is a strategically expedient approach to a fiscal crisis for the region's family planning services. As family planning associations are no longer required to meet a national imperative of population control and as funds become increasingly scarce for family planning services, HIV/AIDS emerges as an important avenue of securing local, regional, and international funds, particularly because of the aforementioned endorsement provided by the MDGs.[19] This, however, further narrows the possibilities by which women's reproductive options are imagined and lobbied for by one of the few advocacy units in the region. The coupling of reproductive wellness with the war on HIV/AIDS facilitates a more integrated interpretation of MDGs Three, Four, and Five in relation to Goal Six's objective to halt and reverse the spread of HIV/AIDS by 2015.

This expedient coupling of reproduction with HIV makes strategic sense in light of the uni-dimensional policy framework that is applied to the region—HIV/AIDs becomes the pathway to funding relevance. For example, the PAHO guidelines for the implementation of MDG Six, which notes that eight of the eleven high-incidence countries in the Latin-American/Caribbean region are based in the Caribbean. PAHO has pointed out that 21 percent of the 2.4 million individuals living with HIV/AIDS reside in the Caribbean, thereby making the region the second most HIV-affected region in the world.[20] Such accounting, though seemingly benign, results in an over-determination of the region as HIV/AIDs infected. Further, this coupling, while strategic, by virtue of its very expediency may prove deleterious to any attempts to mount equally as important arguments for women's autonomy, rights, and pleasure.

It is for these reasons that I worry that the MDGs, once integrated into the Anglophone Caribbean, will do little to advance the political project of Caribbean women's equality. Philip Alston argues that it is an optimistic premise to believe that the MDGs will easily translate into a discussion of rights. Alston notes that the "barebones version that is sometimes put forward might accord only a token role to civil and political rights and endorse a very limited portion of the overall social and cultural rights agenda" (2005, 760). It is disturbing that, in the face of this incapacity to address philosophical questions of entitlement, being, and rights, the MDGs are rapidly and enthusiastically being operationalized by state-managers in the region with little care to correct for significant gender omissions.

Besides the conceptual limitations of the MDGs, it is equally important to note the inability of previous organizational models for achieving gender equity (e.g., any gender mainstreaming model) to challenge the limitations of the MDGs. The gender mainstreaming model, which has served as the primary way of incorporating women's concerns into the national machinery, is ideologically and structurally unable to redress the MDGs' limitations. However, feminist advocates should see the MDGs as an opportunity to engage the state and hold it accountable for women's embodied erasures from contemporary development models.

FINDING A WOMB OF ONE'S OWN: REPRODUCTIVE ADVOCACY IN THE ANGLOPHONE CARIBBEAN

Undoubtedly, family planning approaches remain critically important for small island-states with limited financial resources. The historical tendency toward a primarily family planning approach is governed by what anthropologist Kalpana Ram refers to as a disembodied rationality (2001, 101).[21] In this sense, family planning becomes a technical demographic exercise in which planning is "applied directly on a body that is conceived only as the passive carrier of a discrete biological fertility" (ibid., 101). The technical-rational aspect of such an instrumental approach effectively forecloses any broadening discussion about women's intrinsic value as agents. This has two consequences: the first is that it severely limits our understanding of what constitutes a "choice" for women. Consequently, "choice" under these parameters circumscribes women to choosing from the array of possibilities that are compatible with the state's agenda (e.g., small family size). The corollary of this, however, is that women's voices are delegitimized from all connected areas of concern (e.g., the desire for non-procreative sex).

This technocratic approach has a long history in the region. Trinidad and Tobago's post-independence era saw a strong and supportive alliance between government policy and family planning services, both finding common ground in the desire for population control. This alliance was marked by a rapid increase of family planning clinics, a spike on contraceptive use by women, and an abeyance of Catholic opposition. Legal scholar Dorothy Roberts, citing the work of Trinidadian demographer Jack Harewood, noted that between 1960–1962 the FPA saw 450 new clients; seven years later, the roster of new clients had risen to a remarkable 15,620 (2003, 11).

In so far as we might name the "rights" issue of the day as women's access to contraceptives, we might say that there was significant political support for an expansion of women's reproductive choices and women's right to access contraceptives. In reality, though, early state involvement in women's reproductive lives was inextricably intertwined with nationalist sentiment and macro-planning (Roberts). For example, the growth in infrastructure and services during Trinidad and Tobago's early post-independence offered women increased control over their reproductive lives. The danger of securing a broader range of reproductive options through the primacy of a population planning approach, as with all instrumental approaches, is that women risk losing these rights or being hindered from expanding them further when these gains are no longer in alignment with the *state*'s overall objective.

The historical antecedent of reproductive rights in the Anglophone Caribbean is a peculiar mix of confrontation and concession. The 1970s and 1980s might best be seen as the region's era of confrontational body politics. Here, Rhoda Reddock chronicles a number of important advances during this period, including the Maternity Leave Law Campaign in Jamaica (1978–1979), the Campaign against Violence to Women in St. Vincent and the Grenadines (1985–1986), Trinidad and Tobago's Sexual Offences Bill (1985–1986), and the Jamaica Association for the Repeal of Abortion Laws (Reddock 1998, 60–61). Where broad abortion law reform occurred, what we find is a form of state-led concession: namely, concession to address the contradictions of criminality, which, because of the collusion of all parties involved, protects doctors and/or backstreet abortionists while leaving economically disadvantaged women at medical risk; concession to years of lobbying and advocacy; and concession to the economic costs of criminalized abortion. Concession to abortion law reform in the Caribbean occurred prior to the transnational shift from population control to reproductive rights signaled by Cairo in 1994.[22] Ironically, contemporary reproductive rights advocates find the political environment less than responsive to arguments for women's reproductive autonomy. This unresponsiveness is exacerbated by the lack of support that reproductive rights activists find within platforms such as the Cairo and Beijing platforms and the MDGs.

In the Anglophone Caribbean, feminist NGOs have begun to challenge the assumed "right" of state-managers to be the sole or leading arbiter of women's reproductive capacity. As the lobby continues, it will raise questions about the discriminatory class practices that surround women's access to information (e.g., knowledge of contraceptives), as well as the class differentials that exist in securing safe, albeit illegal, abortions. Poor and underprivileged women remain the constituency most hurt by the anti-choice lobby of the religious right and civil society. The pro-choice lobby foregrounds as well the state's lack of investment in policies pertaining to women's (control of their) reproductive health (e.g., the capacity to enforce the use of contraceptives) and sexual well-being.

At the forefront of this contemporary reproductive rights advocacy is the feminist NGO ASPIRE, started in Trinidad. ASPIRE registered as an NGO in October 2000, with a stated mandate of reproductive health and equity. Currently, they operate with a full-time director, three full-time staff members, and a number of volunteers. Their present focus is on abortion law reform. Their mode of organizing, their efforts at coalition building, and their commitment to research-based advocacy offer important contributions to earlier efforts on improving women's reproductive health in the Caribbean.[23] They frame their arguments for abortion law reform by highlighting the economic costs of unsafe abortions. Choosing to ignore or deny the occurrences of abortion, ASPIRE claims, does not reduce its occurrence nor mitigate its related economic and social costs. Advocates for Safe Parenthood and Reproductive Equity's research suggests that for Trinidad and Tobago, there are possibly as many abortions as there are births, that 3,000–4,000 women are treated at public health facilities due to complications from unsafe abortions, and that such fallout carries an economic cost for the public health sector and others due to women's maternal-related morbidity (Martin 2005).

Roberts, commenting on ASPIRE's work, concluded that their approach uses "a nationalist discourse that frames access to safe abortion as a public health issue, but that places women's interests at the forefront" (2003, 28). Drawing on my earlier use of Sen's distinction between instrumental and intrinsic value, I think it may be more helpful to see ASPIRE's approach as one that is strategically drawing on instrumental language to facilitate their social justice cause. Instrumental arguments—though limited—offer rationally sound arguments. Religious mores may account partially for the lack of persuasiveness that instrumental arguments have carried with state-managers, but it may also be necessary to consider the invisibility of poor women in the region. The stigma attached to abortion makes it less than likely that women who are most vulnerable to the state's failures will publicly protest these failings.

Historically, Caribbean state-managers circumvented any public furor on Caribbean women's reproductive lives through top-down implementation of policies that were infused with a line of instrumental reasoning in which women became another variable toward getting the macro-economics right.[24] Earlier examples of this approach, which took the form of curbing women's reproduction through the provision of contraceptive services to mitigate against the negatives of over-population, are echoed today as feminists attempt to speak to state-managers using the familiar language of instrumentality. It is not that this approach has not been without its successes. As discussed previously, the former has been a highly successful technical approach that, with the decrease in fertility rates in the region, has now aligned itself through health management approaches in the battle against HIV/AIDS. Table 4.1 indicates that the provision of family planning services either remains constant or increases, offering a landscape in which the institutional infrastructure continues to improve while fertility rates decline.

Instrumental arguments must be aligned with the ends that are of interest to those who hold power. To engage in this kind of argumentation as the basis of feminist advocacy will offer gains, but gains that do not necessarily advance society's broader understanding of women's value, worth, and capacity as agents. It is not surprising, therefore, that ASPIRE's arguments regarding maternal morbidity and mortality have been slow to shift the political and legal landscape of the region regarding the decriminalization of abortion.

Table 4.1 Overview of Women's Reproductive Health

Country	Fertility Rates (2003)	Maternal Mortality Rates (per 100,000 live births) 1995 HDR	Measures to Prevent and Manage Unsafe Abortions*	Number of Public Service Delivery Points* ca. 1990	Number of Public Service Delivery Points* ca. 2000
Antigua and Barbuda	2.3	85	Abortion is illegal	23	27
Bahamas	2.3	10	Abortion is illegal except for medical reasons	122	119
Barbados	1.5	33	Medical Termination of Pregnancy Act 1983–1984: Provides for lawful termination of pregnancies.	14	14
Belize	3.1	140	Abortion is illegal; services provided at government hospitals to women who underwent unsafe abortions	70	98
Cuba	1.6	24	Abortion is legal, free and with secure accredited services		978
Dominica	2	—	Abortion is illegal; treatment provided if necessary	52	53
Grenada	2.5	—	Abortion is illegal; treatment provided if necessary	—	—
Guyana	2.3	150	Medical Termination of Pregnancies Act (1995)	—	—
Haiti	3.9	1100	No action taken at government level	—	—
Jamaica	2.3	120	Abortion is illegal; management of unsafe abortion is part of routine obstetric care in hospitals		366
St. Kitts and Nevis	2.4	—	Abortion is illegal; special legislation in case of incest and rape	21	21
St Lucia	2.3	—	Legal under certain conditions	34	—
St. Vincent and the Grenadines	2.2	—	Abortion is illegal; complications treated in hospitals		39
Suriname	2.4	230	Abortion is illegal; complications treated in hospitals	—	—
Trinidad and Tobago	1.6	65	Abortion is illegal; permitted only in cases in which mother's life is at risk		106

Compiled from data taken from UNECLAC Review of the Implementation of the Cairo Programme of Action in the Caribbean and fertility mortality rates from Azziza Ahmed's *UNECLAC Report to Fourth Caribbean Ministerial Meeting on Women*, February 11–17, 2004.

EVERYBODY'S BUSINESS: WOMEN'S REPRODUCTIVE HEALTH AND LAW REFORM IN THE ANGLOPHONE CARIBBEAN

At the point in which women attempt to exercise agency that counters the heteronormative, maternal paradigm promoted by the state machinery, we find a fertile space in which to examine the tropes of identity that circulate around notions of womanhood, nationhood, power, and citizenship. This meeting point of state and maternal body brings to the fore women's dependence on and subordination to the state as the benevolent patriarch, most readily seen in responses to decriminalization of abortion.

Rather than beginning from an absolute premise of illegality, I want to use Table 4.1 to discuss reproductive rights in the context of degrees of permissibility. For example, while Antigua and Barbuda's criminal law penalizes women terminating pregnancies and those providing abortion services, many Caribbean territories permit lawful terminations in varying instances of crisis and harm and at certain stages of the pregnancy (e.g. rape, incest, and health of mother and/or child.). [25] Barbados[26] (1983) and Guyana[27] (1995) offer the broadest legislation on termination[28] in the Caribbean. However, reproductive rights legislation is never merely about a woman's access to termination. Reproductive equity legislation codifies a complex set of political and philosophical reckonings. Abortion law reform dislodges a woman's body from a very peculiar kind of state ownership, one that narrates its legitimacy through a form of "benign" paternalism bent on protecting women from themselves.

As such, the rigid binary of legal/illegal does not aid a cultural understanding of whether women think abortion should be legalized in their respective territories. Guyana, which eventually secured termination legislation in 1995, did so after a halting twenty-four-year process beginning in 1971, at which point the People's National Congress appointed a Special Select Committee to review reproductive rights law reform (see Figure 4.2). The long process might be best described as a relay, characterized by renewed and discontinued dialogue, changing actors and advocates, public sensitization, community and special interest dialogue, advance and appeasement.[29]

The politics of absolutes are invested in casting rights, resources, and access in terms that disavow and suppress the fluidity of engagement and dialogue. Early data in the Guyanese context suggest that a kind of "no but yes" perception on the legalization of abortion dominates. My use of "no but yes" should not suggest ambivalence; rather, it reflects the dissonance between what women say and do as well as their contextual, contingent relationship to law reform. In the bid toward decriminalization of abortion in Guyana, a questionnaire administered to 481 people (60 percent female) showed that 63 percent of the women had a family member who had had an abortion and 30 percent of them admitted to having had an abortion themselves. An overwhelming 89 percent indicated that despite the "illegality," this procedure (theirs as well as the relative's) was done by a medical practitioner. Yet, many medical practitioners (nurses, pharmacists, doctors), when asked if "the law on abortion should be liberalized," said no. Despite the proclamation of illegality, Table 4.1 reminds us that there are a number of conditions (medical reasons) under which doctors have been given freedom to perform abortions in the region. However, above and beyond such permissibility, enforcing these codes of criminality is difficult largely because all parties have a vested interest in protecting the other such that existing codes of criminality become a non-law due to their lack of enforceability.[30]

The contingent nuances of abortion law reform were further evident in women's responses to law reform. When the medical records of 1,000 women were examined in Guyana's capital, Georgetown, almost half of all the women reported having had an abortion (48 percent). When asked more contextual questions (e.g., about a woman's capacity, health, rape, incest), the responses shifted to an affirmative. Fifty percent of high school students said they objected to law reform but then listed an affirmative response when presented with specific scenarios (Nunes and Delph 1997).

This point of departure allows for a more subtle discussion, which reminds us that many regional practitioners do perform terminations both in and out of the context of crisis (Roberts 2003). The prevalence of abortion with or without law reform is itself the strongest argument for law reform, and the strategies toward law reform should avoid the binary of il/legality to better reflect an understanding of the concrete scenarios that fill women with dread and worry regarding their reproductive well-being. Avoiding the binary of il/legality prompts us to think in more precise terms about what is at stake when feminist advocates demand an expanded platform of reproductive rights which extend beyond contingencies of danger, risk, or health.[31] I have attempted to argue that comprehensive law reform directs attention to women's intrinsic value as opposed to only facilitating material gains to the state. It positions women's right to safe delivery services as critical to a reduction of maternal mortality and acknowledges women's rationality and their ability to make decisions about their reproductive well-being and sexual pleasure. Yet, social convention not only suggests to us but allows us to treat women who conceive an unwanted pregnancy as suspect in character for its very occurrence.

The socio-political context of reproductive rights in the Caribbean is further complicated by the levels of race and class collusion that makes advocacy difficult for all stakeholders. There is an understudied space of class silence and complicity between the women who campaign within the pro-life movement and the intimacies, for example marital and filial ties, they share with medical practitioners who actually perform abortions. In addition to these class and familial connections, the medical profession is still very much a revered profession in the region. As such, class reverence extends a degree of societal protection for medical practitioners. This was ironically evident in the Guyanese context where pro-life protesters were encouraged by their North American counterparts to adopt a very U.S. style of lobbying and began to picket the offices of doctors known to perform abortions. Regaled with placards and impaled dolls, the perceived "inappropriateness" of picketing doctors' offices rather than harnessing Pro-Life support, did quite the opposite and produced significant social ire (Nunes and Delph 1995, 15).

Family planning approaches maintain a long history in the region, with Barbados as a forerunner, establishing its family planning association as early as 1955. Further, regional law reform precedes Cairo, not only with the passage of legislation to secure terminations in Barbados in 1983, but also with the 1974 Caribbean Common Market Health Ministers resolution that regional legislation be reviewed with the aim of improving maternal health (Nunes and Delph 1995, 12). Despite the limitations that Cairo places on feminist reproductive rights advocacy, the ICPD, like Beijing, has been repeatedly endorsed by Caribbean states. Following adoption, the Cairo Program of Action was reaffirmed in 2003, in Port of Spain, Trinidad, at the Caribbean's Sub-regional Meeting.

The fact that the Caribbean region has not only repeatedly affirmed their commitment to Cairo and Beijing but also that discussions of abortion law reform long preceded such discussions transnationally presents an additional layer of contradiction to contemporary paternalism. Not without consequences, this paternalism—characterized by the invisibility of poor women and their maternal vulnerabilities, by platform affirmation without operationalization, and by strident resistance to feminist advocacy—ensures that Caribbean women's health is yet to be comprehensively incorporated into the state's policy and planning framework. Abortion law reform requires a different kind of engagement and concession, namely, a willingness to cede male privilege over women's reproductive possibilities. Law reform which has been secured in the region has been so socio-politically idiosyncratic that it is difficult to crystallize any transferable political strategies. However, despite the appearance of a top-down imposition of decriminalization legislation, a closer look shows that community and special interest dialogue over a very long period of discontinuous advocacy are emerging commonalities in the two territories where broad-based law reform has occurred.

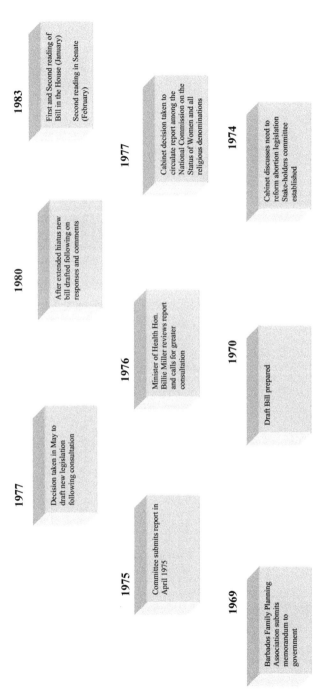

Figure 4.1 Timeline: Passage of Barbados Medical Termination of Pregnancy Act, 1983 (constructed from Barbados' House Debates on Medical Termination Draft Bill, 1983).

118 *Feminist Advocacy and Gender Equity in the Anglophone Caribbean*

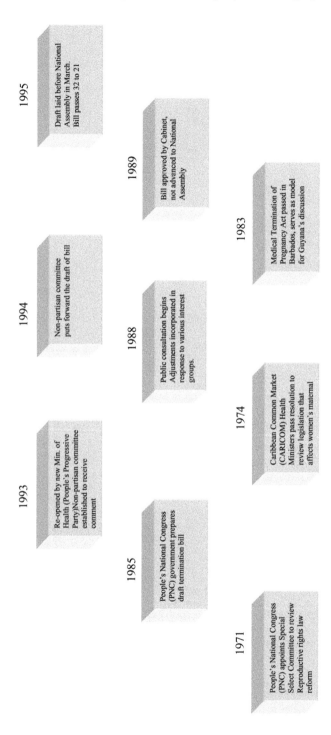

Figure 4.2 Timeline: Passage of Guyana Medical Termination of Pregnancy Act, 1995 (Act. No. 7 of 1995; constructed from "Making Abortion Law Reform Happen in Guyana: A Success Story," Nunes and Delph, Nov. 1995).

Barbados has remained both a pioneer and an anomaly in the passage of legislation to secure termination from as early as 1983: a pioneer because its legislation far preceded Cairo's articulation of its goals in 1994 and an anomaly because of the virulent resistance that the country has taken on similarly debatable issues, such as the 2002 national debates on homosexuality and on the decriminalization of sex trade work. The latter comes as a reminder that there is still a degree of unpredictability to expanding the terrain of human possibilities. As such, we cannot assume that strategies are easily transposed from one area of advocacy to the other. Each terrain of advocacy demands that it be fought on its own contextual, contingent, and intersectional terms.

The need for long-term strategizing is apparent in the case of both Barbados and Guyana. Barbados's abortion law reform was revived and spearheaded in its final leg by Dame Billie Miller, the first woman to sit in the Barbados cabinet.[32] Very early in her capacity as Minister of Health and National Insurance, Miller attempted to introduce abortion law reform but recalls that she was dissuaded from doing so for fear of electoral repercussions. In turn, she engaged in a six-year lobbying effort in which she deployed her personal capital to garner support, to build coalitions, and through dialogue to appease potential opposition before bringing law reform legislation before the Barbadian House of Assembly on January 11, 1983. Under the purview of her unrelated portfolio as Minister of Education and Culture, Miller successfully introduced the bill and in the second reading of the bill lamented that in 1983 ". . . women find themselves in Barbados in this situation where in no other medical condition are they so humiliated. They have to hide under cover of night and other cover" (House of Assembly Debates 1983, 2574–2575).

Miller identifies her early family planning work as a political catalyst for her commitment to abortion law reform. Allies to the passage of this legislation similarly identified their work in communities and personal knowledge of young women who encountered complications from unsafe abortions as one of the most persuasive factors toward their supporting reproductive rights legislation in Barbados.[33] However, Barbados's discussion regarding law reform began as early as 1974. During the second reading of the bill, then Prime Minister John Michael Geoffrey Manningham "Tom" Adams noted the genesis of the present legislation as May 1974 with a Cabinet decision "to examine the need for a reform of the law relating to abortion, with particular reference to the legal, sociological, economic and religious aspects thereof, and to make recommendations thereon" (House of Assembly Debates 1983, 2594).[34] Adams described his support of the bill as "the only honest and moral thing to do" (ibid., 2596).

Much of the debate in both the House and Senate turned on whether the legislation was an endorsement of "abortion on demand." Leader of the Senate and supporter of the bill, Senator Barrow, declared it to be a "compromise bill," arguing:

> We have gone through all the consultation possible, we have come right down to the wire, where we realize that if we take the position as advocated by the Family Planning Association, if we take the position advocated by the Status of Women Commission, if we take the position mandated by any reasonable interpretation of the statistics on abortion, then what this Bill would have to provide for is abortion on demand. On the other hand, if we take the sensitivities of ordinary right thinking Barbadians into consideration, if we took the views expressed by the various churches into consideration, then we come with a Bill which does not in any way represent abortion on demand, but which says that, on grounds of humanity, there are certain cases in which to perform an abortion should not be criminal. (Senate Debates, February 1983, 901)

Abortion law reform extends beyond the acquisition of new legislation. In so far as law adjudicates sex and sexuality, law reform also signals a profound concession of male privilege and patrimony over women's bodies. Abortion law reform, therefore, expands women's access to the terrain of citizenship in ways, as I have suggested earlier, that allows them to participate in decisions about their own bodies and in the national project as law-abiding citizens rather than to be excised as criminals. Nitza Berkovitch asserts that "the concept of citizenship implies participation in public institutions and a notion of personhood in the legal and bureaucratic sense" (1999, 10). Within this attempt to clarify the content of "citizenship" lies the ongoing need to confront those obstacles that marginalize women's participation and, by extension, constrain their access to what Berkovitch refers to as "personhood." Abortion law reform engages and corrects for the ways in which women in the Caribbean have been historically excluded from participating in legal and bureaucratic decision making over their reproductive well-being. It corrects for state-mandated expressions of citizenship predicated on women's invisibility and silence, not only from the state, but from themselves and their bodies.

The initial statistical upswing that may follow abortion law reform can be accounted for by more accurate data gathering, and there is growing consensus that abortion law reform does not lead to an increase in abortions and it certainly allows for safer abortions. Tables 4.2 and 4.3 show that for both Guyana and Barbados, termination rates have held steady or declined, a challenge to the popular belief globally and in the Caribbean that abortion law reform would lead to a spike in the prevalence of abortion. However, because pre-legislation data is at best inadequate and at worst absent, no comparisons can be made.

Table 4.2 Abortions Rates at Queen Elizabeth Hospital, Barbados (1991–2001)

Year	No. of Abortions
1991	709
1992	723
1993	593
1994	588
1995	484
1996	533
1997	583
1998	644
1999	526
2000	528
2001	645

Adapted from *Maternal Mortality, Abortion and Health Sector Reform in Four Caribbean Countries* (Ahmed, 2005).

Table 4.3 Abortion Rates, Guyana (1996–1999)[1]

Year	No. of Abortions
1996	7,531
1997	6,614
1998	5,591
1999	3,310

[1] These figures represent data submitted by doctors to the Abortion Advisory Board and may potentially reflect some under-reporting. The figure for 1996 represents a ten month period of reporting (March–December) (*Guyana Chronicle* October 23, 1999). The difficulty experienced in trying to compile these data again drives home an additional danger of coupling women's reproductive health with HIV/AIDS. We gather data for issues that matter to us; more importantly, the data we gather, in turn, shape how we understand women's reproductive value, which increasingly is being shaped statistically in terms of what it can tell us about HIV/AIDS rather than about women's reproductive autonomy.

Several concerns arise out of the Guyanese and Barbadian context. To what extent are women aware that the procedure has been legalized? To what extent do the stigmas which may still be associated with securing an abortion serve to dissuade women from using formal services and subjecting themselves to the dangers of unsanitary backstreet abortions? What impact do declining health services hold for women who may wish to secure abortions formally? How have training procedures of medical practitioners and support staff been adjusted to facilitate the medical procedure? I flag these questions in light of the continued reports of Guyanese women dying due to complications arising from unsafe abortions during the post-legislative period. These questions further show the limitation of an instrumental rationale toward formulating legislation for women and other minorities. Instrumental approaches to law reform make the passage of legislation the accomplishment. Matters that hold significance for women's intrinsic value, such as public health delivery services, medical training, appropriate reporting, and community education, are disregarded since the state's vulnerability has been secured. In Barbados, for example, it remains difficult to secure any reliable data for maternal mortality due to abortions because of under-reporting by doctors within the private sector (Nunes and Delph 1995). Instrumental discussions side-stepping women's worth make it easier to foreclose on or sustain a discussion that focuses on what women are able to do with their access to legislation.

Women's reproductive rights in a post-Cairo and post-Beijing period, now supplanted as it were by the MDGs, present a recalcitrant landscape for women's reproductive health. Limitations notwithstanding, neither the adoption nor operationalization of these platforms in the region occurs with women's reproductive well-being in mind, but instead, a further excision of women's reproductive options from mainstreaming discussions within state machinery. So, having examined the process of law reform in Barbados and Guyana in the pre-Cairo era, it might be useful to examine the irony of state resistance in the post-Cairo and post-Beijing era. The unfolding situation in Trinidad and Tobago represents an attempt to secure the passage of legislation out of a feminist lobby/advocacy approach. Trinidad and Tobago, then, stands as an important case study of contemporary feminist advocacy challenging the state's hold on women's bodies and sexual agency.

While maternal mortality remains low, complications from unsafe abortions stand as a lead cause for women's morbidity (Ahmed 2005). Clearly cognizant of the relationship between national health concerns and provision of social goods, ASPIRE articulates its long-term vision as one that aims to secure improved reproductive health. Interestingly, despite their use of research data showing the widespread occurrence of abortions in Trinidad and Tobago and the cost of these unsafe terminations to the state, both in terms of women's health and the financial output by health institutions, ASPIRE has not had adequate success in securing the support of the relevant Ministerial authorities (e.g., Ministry of Health, Attorney General, Prime Minister's Office, or Gender Affairs Division). The lack of political response by Trinidad and Tobago's state officials is particularly stark in light of the fact that ASPIRE's research documented that approximately US$576,700 per annum is being spent on health care for complications arising for women who have undergone unsafe terminations and that, annually, approximately 4,000 women require additional medical care due to complications from unsafe abortions. In the midst of the debate, then Prime Minister Patrick Manning declared that his was a pro-life government and, later, with jocular irresponsibility, admonished the residents of a working class community to watch more television for its contraceptive effect (*Trinidad Express* August 17, 2005).

Here, we begin to see the limitations of mainstreaming models around the question of reproductive care and well-being. State-managers can be intractably unwilling to acknowledge women's value even when the cost is assigned in stark neo-liberal efficiency or cost opportunity terms, and feminist advocates are left with little by way of international platforms to facilitate moral suasion.

It does not require extensive argumentation to convince one that Caribbean officials have not exhibited the political will to address women's intrinsic value or to redress women's ideological disadvantage with society. This is especially so with reproductive health and well-being as developmental issues. Both structurally and in an embodied sense, the political system is so governed by partisan loyalties that it is difficult for independent feminist advocates to exist effectively without mainstream political endorsement. However, coupled with this lack of political incapacity is the embedded danger that the political position of the individual becomes the im/possibilities of a wider, more varied constituency of women.

While I've examined the political and bureaucratic resistance toward achieving reproductive dimensions of embodied equity, I want to examine the other constraint on achieving equity in the region: the disassociation

Table 4.4 Maternal Mortality Rates, Trinidad and Tobago (1991–2001)

Year	Maternal Mortality
1991	49.18
1992	60.70
1993	66.40
1994	76.2
1995	67.5
1996	38.9
1997	70.4
1998	44.7
1999	38.2
2001	70.4

Source: Aziza Ahmed/Dawn Steering Committee, Adapted from Republic of Trinidad & Tobago, Central Statistical Office, Population and Vital Statistics Report, 1999 and CEDAW Report, 91.

between mainstreaming and the feminist politics of the agent charged with responsibility for that mainstreaming.

In this regard, the St. Lucian context highlights one of the ways in which women's issues become caught in the institutional quagmire of apolitical system. In September 2004, the St. Lucian government, headed by Kenny Anthony, introduced Clause 166 of the Criminal Code. This revised the existing law to allow St. Lucian women access to an abortion under certain conditions, i.e., rape, incest, or the health of the mother. This in practice made St. Lucia no different from St. Kitts and Nevis, Jamaica (broad health grounds), or Trinidad and Tobago (mother's life at risk). For all practicality, unless a woman could *prove* (with a police report) that she had been subject to one of these unfortunate scenarios, securing an abortion remained illegal.

One would not think this to be the case, though, with the furor that subsequently unfolded in this predominantly Catholic population.[35] On September 16, 2004, the bill was read in the House of Representatives. Ten days later, approximately 2,500 citizens protested plans to amend the code that, for all intents and purposes, did not aim to legalize terminations. At the forefront of the march was a young female Minister of Government responsible for the Ministry of Gender and Home Affairs, Sarah Flood-Beaubrun. She was pregnant, and she opposed the Bill. During the debate of the bill, the then Minister with responsibility for Gender and Home Affairs Flood-Beaubrun voiced her dissenting view by calling the Members of Parliament "murderers" and "child killers." It was reported that, in her opposing speech, she used the terms "murderers" and "child-killers" no less than 32 times (*The Star* January 23, 2003). She was subsequently dismissed (January 2004) and has since resigned from the St. Lucia Labour Party (March 24, 2004).

The St. Lucia case, like Trinidad and Tobago's, presents clearly the ways in which critical gender mainstreaming concerns are often locked in a quagmire of state politics. However, St. Lucia's case introduces an additional complexity, where the possibility of mainstreaming reproductive equity for women is blocked by the deliberate opposition from the person with central responsibility for such processes. A juxtaposition of Miller's and Beaubraun's work toward the passage of reproductive rights legislation prompts us to revisit the transformative potential of effective "femocrats." Eisenstein's discussion of femocrats as women who are powerful "in government administration, with an ideological and political commitment to feminism" can only be cautiously applied to the Caribbean because there it is not a given that these women will have an ideological commitment to feminism or that they will represent power within government. In addition to this, the two scenarios suggest the need to question the logic of mainstreaming, which as yet does not adequately call for a policy and planning distinction between a femocrat's ethical and moral locations. St Lucia's case, therefore, highlights the ways in which gender policy can go awry in the absence of a negotiated dialogue delineating an acceptable range of desirable capabilities for women.

RE-GENDERING THE YEAR 2020 VIA THE MILLENNIUM DEVELOPMENT GOALS

Jacqui True cautions that the agenda for the twenty-first century is not the avoidance of co-optation, but strategies of negotiation, is a pertinent one. However, I attempted to argue that in the area of reproductive rights, reproductive rights strategies of negotiation from within the state only appear feasible if femocrats carry an inordinate amount of power and credibility within their respective grouping. Further, I questioned the tendency within the region to lobby for abortion law reform using an instrumental efficiency model. Although this model is certainly an effective strategic one, to marginalize discussions about women's intrinsic value may limit the implementation possibilities and long-term effectiveness of the goals secured.

Development brings with it its own benign hegemonies and invisibilities, and the MDGs are no exception regarding women's reproductive rights. The conciliatory positions within ratified and adopted platforms, such as Cairo and Beijing and CEDAW, increase the local lobby toward law reform because of the ways in which these documents cede ground to national law. Further, in the growing embrace of the hierarchy of developmental needs, the deepening ties that exist between reproductive equity and the war against HIV/AIDS complicate the extent to which advocates are able to address reproductive autonomy. I maintain that these inscriptions cannot be adequately challenged by existing mainstreaming approaches, first, because gender mainstreaming in the Anglophone Caribbean is too often driven by expediency and instrumentalism within hostile bureaucratic environments and, second, because these mechanisms are too frequently charged to change agents who have a limited understanding of the (ramifications of) stratagems required to achieve gender equity in the Caribbean.

The MDGs' objectives now guide the strategic management plans of a number of countries worldwide. This embrace lends a degree of urgency for a philosophical re-fashioning of questions, such as What is meant by development in the Caribbean context? and Development for whom? Pushing beyond material access and needs, what, then, is the nature of gendered humanity that will govern social citizenship in the twenty-first century?

Women's reproductive health and sexual well-being are closely tied to the question of self-determination and citizenship. Within present developmentalism, women's self-determination cannot be confined to a fiscal argument of resource allocation for health care and family planning. This formulation becomes one whereby women serve as a means to the state's end rather than the state-machinery enabling women's ends. We cannot dismiss the analytical importance of the socio-political and psychological landscape that facilitates women's choices. Terms such as "human," "citizenship," and "development" are critically implicated in the impasse around women's reproductive rights in the Caribbean. Consequently, any discussion of women and development must be informed by a discursive reading of women's bodies for the several subtexts that have inevitably formed part of their capacity to engage with notions of humanity and citizenship. This corporeal interrogation helps us to notice the ways statistical data, laws, and policy are always at work shaping bodies and, further, how different bodies such as the maternal body are discriminatorily incorporated into the political and legal landscape in ways that limit claims to citizenship.

Achieving Caribbean women's reproductive equity possesses the capacity to substantively transform how the Anglophone Caribbean engages the idea of rights and entitlement, adjust the language we use around discrimination, reform the practices that we enforce, and prompt us to think more reflectively about resources that we invest into securing more equitable societies. In sum, Caribbean women's reproductive equity may actually serve to be the DNA of what our eventual ideas of democracy, equity, and citizenship become.

5   Keeping the Mainstream in Its Place
Sexual Harassment and Gender Equity in the Workplace

I am sitting at the table. I am anxious. My father, after a stint as a carpenter on an American military base, warned his "girl children" to be wary of "men in uniform." My friends, tongue in cheek, would probably say I have been zealous in my adherence to his missive until today. I am waiting to deliver the findings of a recently concluded research project on gender equality in the Royal Barbados Police Force (RBPF). The Senior Command Team enters, but previously warm interviewees do not look at me. Circular sitting formations only work to equally distribute power among feminists and pacifists because, despite the circular formation of our seating arrangements, it is clear where the concentration of power resides in this meeting—it is not with me. This scene is funny only in hindsight. In hindsight, I better understand male intimidation and how easy it is to be taught how to toe the line without any need for a verbal articulation of where the line is positioned. Once more, I wonder at the implications of such an environment for marginalized bodies and subjectivities. In hindsight as I review the research data, I wonder why as Caribbean feminist scholars we have been so slow to analyze the connections between mainstreaming women into atypical professions as a mode of equity and the potential backlash effect of this mainstreaming—sexual harassment.

A recurring theme throughout *Feminist Advocacy* has been the analytical inadequacy of realizing a critical mass without commensurate attention to redressing cultures of inequity. Two caveats are needed here: first, that pursuing a critical mass remains an important aspect of redressing inequity and, second, that in the absence of having a critical mass for many positions of authority, we are yet to know the potential effects of this on existing discriminatory cultures. Pursuing a critical mass must work in tandem with policies that contest institutional cultures of inequity. Without this two-prong approach, attempts to achieve equity by "mainstreaming" marginalized subjects into all sectors of society may feel like being thrust into an institutional minefield, leaving those "mainstreamed" without any corrective recourse. Political scientist Carol Bacchi frames this balance as follows, when she writes:

> Despite the claim . . . that mainstreaming works to change organizations rather than women, we are still working with a model which

accepts the male as benchmark and which identifies different treatment as the problem. Rather, we need to examine the impact of gendered assumptions in creating and reinforcing social hierarchies and in framing lives we may not wish to lead. (2001, 19)

As inferred here, realizing any mode of gender equity requires a sustained challenge to existing masculinized cultures. Bacchi suggests here that seeing "difference" as the problem morphs into naming the marginal constituency as the problem—it is the arrival of the marginalized subject's "difference" that makes policy formulation and interpersonal relations tense. Institutional life, in turn, gets formulated in BD/AD time, "before difference and after difference," with before difference time being memorialized as the glorious days of yore. I want to think about this preoccupation with the inconvenience of difference as the "difference distraction." In other words, this concern for the inconvenience of difference turns our attention away from the very gender ideology that makes difference(s) matter and normalizes the practices that aim to keep difference in place. Among these is sexual harassment.

In this chapter, I briefly explore how this "difference distraction" works through the prism of gender to undermine equity in gender atypical work environments. My analysis centers, among other things, the rhetorical strategies, the in/formal reward policies, and the architectural boundaries that work to construct women's difference as the problem that undermines their effective job performance. However, sexual harassment is of central importance to my discussion because of the ways this practice captures the ironic illogic of the difference distraction. On the one hand, women's embodied difference is violently acted on (by others) to maintain the status quo of the workplace and, on the other, sexual harassment is declared as a problem that women have brought, serving as a distraction from redressing the status quo of the workplace.

To explore these complexities of equity in the workplace, I begin by broadly examining workplace gender equity within the organizational politics of the Royal Barbados Police Force (RBPF), with a focus on sexual harassment as a specific feature of gender equity within workplaces that are being mainstreamed. I further contextualize this organizational study within regional efforts to pass sexual harassment legislation. My aim in this chapter is to challenge the practice of governmental rhetoric around gender mainstreaming as a mode of equity, while remaining reticent on the deleterious effects that mainstreaming has on marginalized bodies once "mainstreamed." This silence itself is symptomatic of persistent emphasis on the technical processes of mainstreaming to the exclusion of comprehensive evaluative mechanisms, which should include a discussion of what marginalized constituencies are able to do once "mainstreamed."

GENDERED NATURE OF THE ROYAL BARBADOS POLICE FORCE'S ORGANIZATIONAL CULTURE[1]

Throughout *Feminist Advocacy*, I have explored the ways in which different organizational cultures are gendered and the effect these cultures hold for mainstreaming processes. My concern has been to examine the procedures and practices through which organizations operate as "spaces in which minds are gendered, and men and women are constructed" (Aaltio and Mills 2002). In other words, achieving equity in the workplace requires that we attend to the in/formal rules, policies, and organizational procedures that inform the messages that women and men receive about their proper functions within the organization. To say that an organization has a gendered culture, however, does not tell us how this becomes so, nor does it tell us how this culture becomes embedded, produced, and reproduced on a daily basis as part of one's "normal duty." This, I have argued, requires procedures of historicization and an eye for the contemporary patterns by which gender inequity is reproduced.

Industrial relations theorists David Wicks and Pat Bradshaw mark the importance of organizational gender cultures, seeing them as processes that "create systems of relative advantage/disadvantage, agency/constraint, and autonomy/dependence in terms of a distinction based on sex and gender"; these processes c/overtly structure the normalization of inequity and the incorporation of inequity into the day to day practices of the workplace (2002, 140). Gender cultures within organizations are, of course, reflective of and connected to the larger societal culture in which the organization exists. The resonance that exists between institutional cultures and commonly held beliefs within the wider society makes it more difficult to question the "commonplace" within the workplace. However, because organizations inhabit a more finite space, they also present a unique opportunity to challenge many of these discriminatory processes.

First, I want to discuss how these commonplace understandings have structured a frame of gender in/equity within the RBPF, with sexual harassment as a particular instantiation of gender inequity within the workplace. The workplace's institutional culture sends two significant messages: first, that men are guardians of the nation and, by extension, embody law enforcement; and, second, that women are, at best, contingency officers with the capacity to undermine the activities of "true" policing activity.

The RBPF recruited its first four female officers in May of 1950. The duties of these pioneering women were governed by the duties and regulations, which were later set out in 1955 as an addendum to Police Act 1908–2. While reflective of the ideas of the period, the terms of their service reflect the RBPF's gender-segregated beginnings. The Police Act Rules of 1955 made pursuant to Police Act 1908–2, in part, read as follows:

130   *Feminist Advocacy and Gender Equity in the Anglophone Caribbean*

>   6. Police women will not sleep in barracks. A roster will be maintained and kept in the charge office, Central Police Station showing the names of the women constables for duty each night. *A duty woman constable shall be required to remain in her home all night to be available for duty if required.*
>   9. Police women are first and foremost police officers and may be employed on any form of police duty in which they may be useful. *They should specialize, however, in work for which they, as women, are best suited.* These duties include:
>   a uniform patrolling
>   b) duties in connection with women and children reported missing or found wandering, destitute or homeless
>   c) duties in connection with girls and children who have been the victims of sex offences or are in moral danger
>   d) taking statements from women and children
>   e) executing warrants upon women and girls
>   f) escort and court duties in connection with women and girls
>   g) searching and fingerprinting women and children
>   h) observance upon women prisoners detained in hospitals
>   i) supervision of school crossings
>   j) plain clothes duties and detective work
>   11. *Police women must exercise their powers of arrest with discretion when dealing with male offenders, securing, if possible, the aid of a male constable.* (emphasis mine; June 8, 1955)

This early history is important to our contemporary understandings of gender equity in the RBPF. Early female recruits were segregated, confined to duties that were feminized, and subjected to paternal protectionism (e.g., staying at home although scheduled for night shift). These rules marked the terms by which women were accepted into the RBPF. Consequently, sex segregation forms part of the institutional memory of the RBPF, and many of these legacies remain today. These sex-segregated legacies include limiting infrastructure to accommodate female police officers, gender stereotyping in task allocation, and positing male officers as the normative agents of the force.

Older and, now, senior officers remember the first cohort of women as idealized matriarchs, a nostalgia that contrasts starkly with the rhetoric used for female police officers today. Many older officers referred to these women as "mentors," "live wires," having a "maternal" touch, or as women who "ruled with an iron fist" and whom male officers feared. This language, however, marks a longstanding pattern of rhetorical tension regarding women in the force: there is nostalgia for the strong, powerful female officers of the past and apathy toward the younger, weak female officers in the present.

This dissonance in the reception of women to the force exemplifies how organizational rhetoric adapts to accommodate non-threatening difference. I want to suggest that this rhetoric resolves itself, not because female officers have changed but because the profession is differently segregated. Senior male officers remember these early female officers as confident while simultaneously rendering them as maternal, waxing nostalgic about the ways these early officers cared for them in their youth. These sentiments have taken hold in the institutional memory and set up standards for contemporary female officers. It is possible, therefore, to talk about this early cohort in masculinized terms of power *because* this cohort was confined to feminized jobs and, as such, posed little or no threat to the male power structure of the force.

Further, a wider Caribbean narrative of the acceptable, if not expected, authoritarian maternal figure also serves to contain this pioneering cohort. Consequently, senior officers remembering these pioneering women's authority make sense of such in terms of "maternal fear," which facilitated their being mentored and working their way through the ranks.

As such, this fear/authority remains persistently femininized; a reflection of domesticated power (read "acceptable") without disrupting any of the roles socially designated for women. In contrast, as the profession becomes differently segregated, female officers are present in larger numbers, have access to a wider range of operational policing options, and are now in direct competition with male officers for promotion. Here, as Fiske and Glick (1995) note, women become "victims of their own success": the better women do, the greater their exposure to a range of "disciplining" practices designed to re-inscribe place.

Women now comprise approximately 14 percent of the RBPF's strength. In 2005, this translated into an absolute figure of approximately 184 female police constables (see Table 5.1). Rank as well as retention is a concern. The average age of female recruits for the period of July 2000–2005 was 23 years, five years younger than the average age of female recruits as special constables. At the time of this study, only 17 percent of the female strength occupy ranks above police constable, a mere 2.5 percent of the force's total strength. Interestingly, the length of female service within the RBPF is very short, with over 50 percent of the female strength located within the category of one to five years of service (Strickland 2005). Such a short tenure could potentially lead to some hesitancy in resource allocation for women. However, the push/pull factors of such a short tenure are best placed, as we will explore, in the context of gender equity in the workplace. As the total number of police women increases, the administrative narratives reflect an increasing anxiety that the RPBF is being feminized—not only as reflected by the increasing number of women on the force but also by the deteriorating conditions of the work and the potential exodus of men from the sphere.

POLICING WOMEN: WORKPLACE EQUITY AND WOMEN'S HYPERVISIBILITY IN THE ROYAL BARBADOS POLICE FORCE

With atypical professions, hypervisibility is not merely an expected occurrence, but one way in which those who are new to the domain are pushed to self-surveillance toward conformity and over-performance. Hypervisibility provides fodder for the "difference distraction." In other words, the "data" that come from such microscopic scrutiny form the bases of differentiated treatment and "evidence" for pre-existing ideological biases. If we look long and hard at the same spaces, we will see why "difference" is, at best, inconvenient. When women are "mainstreamed" into atypical areas of work, the impact and perception of their effects tend to exceed their actual numerical representation. As such, women's presence becomes hypervisible, and, accordingly, women become overly culpable for any problems and deterioration within the atypical profession.

Within the RBPF, women's hypervisibility manifests itself in two ways: (1) the difference between perception and women's actual numerical representation within the force and (2) the degree and nature of culpability that accompanies women's numerical representation. Senior administrators either had a more accurate representation of the percentage of women within the force or chose not to speculate on the numbers. However, for senior administrators, the increasing numbers of women within the RBPF held implications for infrastructural capacity, application of organizational procedures, and the quality of performance. In contrast, the rank and file of the force thought that the percentage of women within the force ranged between 25 and 40 percent. This exceeded by more than twice the actual percentage of women within the force. This overestimation matters because it suggests that the perception of women's presence takes on larger than life proportions to which fellow male officers hinge narratives of incompetence

Table 5.1 Comparison of Male and Female Strength and Attrition (2000–2004)

Year	Establishment	Strength	Male Strength	Female Strength	Female Percentage of Total Strength	Male Discharges (%)	Female Discharges (%)
2000	1328	1209	1103	106	8.76	59 (5.34)	6 (5.66)
2001	1328	1284	1160	124	9.6	45 (3.87)	3 (2.41)
2002	1328	1265	1112	153	12.0	54 (4.85)	5 (3.26)
2003	1328	1299	1123	176	13.5	49 (4.36)	1 (.568)
2004	1328	1300	1116	184	14.1	39 (3.49)	4 (2.17)

Table compiled from data provided by the Human Resources Department of RBPF.

and incapacity. For example, as one senior officer commented on the increasing numbers of female officers:

> In my view, it has impacted in a negative way. If you were to take time to move around, you may very well see that there are more female officers than males, and whether you or I think that a police officer is a police officer, the person out there may not necessarily think so. (interview with member of Senior Command Team, July 12, 2005)[2]

It is statistically impossible, given the gender distribution of the force, for there to be more women than men in any given area (with the exception of clerical work, which I will speak to later). However, this sense of women's "overwhelming numbers" coupled with the narratives of incapacity strive to strategically place women police officers on a treadmill of constant justification of their right to be members of the force. As one female officer with five years on the force observed:

> At the end of the day, you have to give the work to those who want the work. If you have fifteen women applying and five men applying, and out of the five, two qualified? Some women gotta learn to stand up. I mean, really, stand up, don't take anything from anyone, and I mean in terms of chat. Learn the job because as a woman in this police force, you have to work twice as hard as a man, but to me, that is not just the police force. I found that everywhere I've worked. You've had to work twice as hard to get the same . . . the same . . . to get the same treatment. (interview with female police constable July 20, 2005)

If we see gender equity as premised on access and what one is able to do with that access (outcome), then this twin prong of hypervisibility and culpability works to push the outcome component of gender equity further away from women's reach. Interestingly, the implications of this gender differentiation within the workplace resonated throughout a number of the interviews, particularly among senior officers. For example:

> *Rowley:* Are you suggesting that it is the public's perception that she will not enforce . . .
> *Respondent:* . . . that she is weak, that is the public's perception and perhaps with good reason.
> *Rowley:* Why do you say this?
> *Respondent:* Again, it isn't fair to say that, because men will do it, too, men will try get away with not doing their duty. I made the point earlier of trying to avoid the (dangerous) situation. I have had personal experience of being in Bridgetown.

Something developed. Female officer on the spot, rather than make her way to the spot to quell it, she turns in the opposite direction. Now, males do that, too, so I'm not sure what I said earlier on is really true. (interview with member of Senior Command Team July 12, 2005)

Here, the play of hypervisibility and culpability feeds a workplace dynamic of women's supposed incompetence and incapacity. It is unquestionably problematic that a female officer has walked in the opposite direction; it is telling that she is assessed differently for doing so. It is clear that avoidance is not exclusive to women. However, in contrast to how men are assessed, in this narrative her motivation to walk away is her femaleness. Only after some reflection on his own narrative was the senior officer able to realize the extent to which naturalized assumptions about "female inability" were at work in his assessment (*Now, males do that, too, so I'm not sure what I said earlier on is really true*).

As with the narrative above, what follows is an excerpt from an interview with a sergeant, which again reflects the ways in which men's unwillingness to work is acknowledged but rendered invisible, seen but not seen:

We have some females who demand their rights and in fact command it because of their performance. We have some outstanding women, but of course, outstanding, it's only in the minority. So, those ones who are extrovert in their action, they are seen and they are promoted. But the ones who believe that by their actions they will be recognized, that does not work, that is the utopian way. If they hope that someone else sees it, it is never seen. The quiet is never seen. It is the loud and the noisy who are seen. But we have a lot of people in the back who do good work, they do very good work. If you were to look at the work of the women and the men individually, you might be surprised to see that the women outdo the men individually ... But seeing that there are over one thousand men to some less than three hundred women, the men would most often be seen. So, starting at a disadvantage of numbers and then not being aggressive, you are always not going to get your due. (interview with Station Sergeant July 15, 2005)

Despite the general consensus within the RBPF on a general lack of morale and increasing apathy toward job performance, women are the ones who are repeatedly rendered as the primary offenders despite the contradictions and awareness that men are equally as apathetic. In other words, the two-prong dilemma of hypervisibility and culpability is such that when women underperform, they are over-recognized, and to be recognized, they must over-perform.

RATIONING FEAR: PUBLIC SPACES ARE FOR THE BRAVE OF HEART

Organizational logic produces a deeply gendered rationality. An aspect of this rationality is the association of "good policing" with physicality, aggression, operational activities, and street-related encounters, or what one interviewee repeatedly referred to as "hard policing." Within policing, these features not only favor masculinized characteristics but also are embedded in promotional procedures and informal rites of community and camaraderie. Such a rationality infuses a commonplace understanding of what makes a "good police officer" throughout the organization and greases the pathway to upward mobility. It serves to construct an idea of policing that rewards masculinized traits and marginalizes skills that women have been socialized to value.

Against this idealized model of the "good officer," women are seen as a threat to good policing as well as the safety of the "good police officer" (a.k.a. male). Women's supposed propensity for being fearful emerges as one of the foundational narratives that feed the idea that women are a threat to good policing and the safety of good police officers. Narratives of women's fear of "hard policing" pervade the interviews and serve to entrench female officers as inauthentic, or less than competent, police officers.

Not surprisingly, as these narratives circulate, women have begun to comply with them and become victims of their own feminization. This complicity occurs in two primary ways: by their own repetition of the gendered narratives that circulate within the organization and by judging other women as harshly as men do. Alongside narratives of fear, women's perceived inauthenticity as officers is rendered through their infantilization (e.g., female officers referred to as "little girls"); their unwillingness and/or incapacity for functioning efficiently ("women don't wan' work"); and their unwillingness to face danger ("woman 'fraid").

These sentiments mirror the findings of Alissa Trotz's study of the private security industry in Guyana where, she argued, many private industries were disinclined to consider women for firearms training because

> They were prone to hysteria. Paradoxically, this qualified women for sites where "early warning signals" were needed. In other words, female officers are deterrents against potential criminals but ineffective against the actual demonstration of illegal intent, portrayed as unable to handle a gun or cope with situations requiring objective and spur of the moment decisions. (1998, 43)

That "fear" is gendered becomes even more apparent when male police officers were asked to explore their own anxieties about policing. It is not that male officers were not fearful. Rather, their responses suggest that they channeled and masked that fear by masculinized traits such as aggression, physicality, and authority. One male officer observed:

> The first thing that we try to get over is that bravery is not really lack of fear. It is how you deal with fear, and the one thing that you have to get over is that you have to gain control quickly. A lot of the problems emerge because people are not familiar with street and street language, in that if you do not have a grounding on the street, the streets are very fearful to you particularly in ghetto areas. Once you stamp your mark early, people respect you. So, you need to work on the streets. Once you work on the streets, you are not fearful and you need to keep coming on the streets to get accustomed to the culture and language and see how the streets change .... (interview with member of Senior Command Team July 17, 2005)

Similarly, another officer, when asked about his coping strategy in a potentially dangerous situation, responded:

> (Slowly smiles) Aggression ... I'll give you another answer. For the most part, the persons that I work with ... let me just add on that there are some males that I would prefer not to work in some situations with ... Is not every situation that you need to be aggressive in. You can go into a situation that might appear hazardous and aggression might just make it worse. (interview with detective July 17, 2005)

Such comments suggest that the institutional discourse of women's fear, far from being innocuous or an individual possession, arises as a *consequence of* the RBPF's gendered culture.

In both responses, for men, fear, once filtered through aggression or authority, can transfigure into a manifestation of bravery. Whereas individual men are fearful, this feature of the job is not discussed as a defining feature of all men. Individual women's fear, however, is, and further, it becomes cowardice and marked as a group identity. As a result, women who manifest courage become exceptional.

The unpredictability and dangers of policing are reasons enough to produce caution and, at times, fear among those who serve. For female police officers, however, the phenomenon of fear sits at the intersection of being a target of aggression themselves, vulnerable as an officer and as a woman. Female officers associated night shifts and street patrols in general with a heightened potential violence and risk to their personal safety. They were cognizant that they were vulnerable *as women* to the same possibilities of violence and assault to which all women are vulnerable. This awareness pertained to both off- and on-duty scenarios, as the following response indicates:

> The uniform gives me a sense of security. I remember going home one night, and they did not have a vehicle and I was afraid. I could walk at 10 o'clock on patrol, but not without that uniform. I had $1.50 in one hand and I.D. in the next, no jewelry and flat shoes. One time, people in Barbados feared police officers. Police come in, people step aside. Now, you have a society where people respect no one. And you have women police out there, and the men, well, they don't even respect the men, they not afraid of them. So what is going to happen with the females? . . . Basic patrol in the city you have to work alone! Come on, this isn't 1860 where people saw you and fled. If something happens, you're alone . . . It doesn't happen overseas. So why should it happen here?! (interview with female police constable July 18, 2005)

This response of fear is not an irrational act of hysteria as suggested within the dominant narratives of women's fear within the RBPF: it reflects a pragmatic recognition of women's general vulnerability to the violence and abuse that women experience in the wider society.

In response to the perception that women "lacked confidence in the execution of their duties," the RBPF initiated a series of self-defense training programs. This program training began on November 24, 2003, and at the point of my field research, the most recent cycle concluded on May 27, 2005. The sessions offered intensive training in unarmed combat for a period of three months, for one hour three times a week. The RBPF required only female officers to attend this training. If the female officers wished to continue at the end of the designated training period, they would have to absorb full costs.

In interviews, female officers who participated in the self-defense training program saw it as enhancing their ability to execute their duties. They felt that the training built their sense of alertness, increased their capacity to defend themselves if attacked, and taught them how to defend themselves without applying lethal force. Two female officers reported that they subsequently had the opportunity to use their training. The following outlines one such scenario of a trained female officer and an aggressive male:

> . . . he became aggressive, refusing to give his name. He was searched. We discovered a knife. So, we told him that since he didn't want to talk to us out there and we found a knife on him, it's an offence and we can take him to the station. We had a struggle, and eventually, because we were three women to a man, rather than being out there and let that person turn you out, I was able to apply a lock on him to take away some of his strength so by the time back up came, we had everything under control. (interview with female police constable July 18, 2005)

One supervising officer, when asked whether he thought the training had an impact on women's performance, responded:

The truth is, no. The ones who were active before are still active and the ones who were inactive remained inactive. I don't know how many classes the inactive ones attended, but the ones who were robust went after it. (interview with Police Inspector July 19, 2005)

The above officer's lack of enthusiasm is not unwarranted. Of 183 females then enlisted in the RBPF, ninety-four (51 percent) had been trained, and sixty-eight (37 percent) successfully completed the program. The low participation and pass rates served to reinforce existing sentiment about women's "unwillingness" to work and were seen as women's active resistance to equipping themselves for the operational aspects of policing. However, a look at the gender differentials of the organizational culture is again helpful. All female officers—and only female officers—were required to participate in the training program. In the words of one supervising officer, "I ordered them all to go . . . the memo from Division Command is "go," but women found a lot of excuses." None of the male officers were required or expected to attend self-defense training, a disparity that was succinctly captured by another senior officer:

The view is, as far as I'm aware, that they were trying to boost the confidence of women, but there is the view and this is our hubris as men . . . It isn't that we lack confidence, it is that we are too confident. (interview with member of Senior Command Team July 12, 2005)

When asked, however, female officers indicated that they were given no time off to pursue training. Junior officers complained that they were expected to attend training after having worked an entire shift:

Who was given time off (with disbelief)? The only time off you were given is if you were working three to ten, and the training was four to five. If you were working morning, you were expected to work from seven to three and then go to training. If you working ten o' clock, you were expected to go and then work ten to seven. It's not time off. That's what they need to understand. (interview with female police constable July 18, 2005)

Female officers' domestic responsibilities after their shift contributed in no small part to the low participation and attendance rates. Further, the training was seen as too short-term to be either adequate or sustainable and additional training was not subsidized.

Both the organizational naming of the problem (lack of confidence and fear) and the perceived remedy (self-defense training) served to further embed women as subpar. Given the gender discourses at work, by singularly identifying female officers to attend additional self-defense training, the RBPF held women to a different standard or marked them as not having skills that men either already possessed or had innately. Female officers saw themselves as being given a heavier work burden than their male counterparts to address a problem that could easily exist among the rank and file of male officers.

Law enforcement's validation of aggression and physical strength places value on certain skills and deemphasizes others. Unless the discourses that undergird such valuation are interrogated, then any action plan with the declared intent of providing opportunities for women and other gender minorities in the workplace may unwittingly perpetuate the very discourses that marginalize gender minorities in the workplace.

I have shown how organizational cultures can reflect and enable gender ideologies at work socially in ways that may be detrimental for women in the workplace, particularly for women who enter non-traditional occupations. The imagined repercussions from gender minorities' impact on the morale of policing undergird the use of hypervisibility and culpability as marginalizing and differentiating mechanisms. Among some male officers, there was the sentiment that women's participation in the police force had already prompted an exodus of male officers. In contrast to this imagined flight, Table 5.1 shows no exodus of men from the police force. Indeed, between the period of 2000–2004, men's attrition actually declined from 5.34 percent to 3.49 percent. Similarly, for women, attrition also declined from 5.66 percent to 2.1 percent. Contrary to the popular belief that women are more inclined to enter the police force primarily to support families, women *and* men indicated that they considered the vocation because it provided a "secure" or "steady" income. Moreover, the allure of policing was not the acquisition of social authority. Rather, both women and men chose policing for its vocational stability, defined in terms of a "steady salary," "good benefits," and "difficulty in getting fired."

The "difference distraction," an organizational red herring, can result in reductionist responses to very legitimate concerns. For example, the growing concern about low levels of male application/recruitment into the RBPF is overshadowed by organizational worry regarding the number of female applicants rather than finding explanations that might account for the declining number of male applicants. As with the over-estimation regarding women's presence in the force, there is similarly the "popular" belief that there is approximately a 10:1 female/male ratio of applicants to the force, which again reflects women's hypervisibility within the force.

As Table 5.2 indicates, contrary to popular sentiment, the female/male applicant ratio remained fairly equal, with the highest disparity averaging 5:3 in 2004. Despite the slightly higher number of female applicants to the RBPF, it is noteworthy that they are actually recruited in *proportionally* smaller numbers than men.

Gender inequity can appear in decidedly rational ways. For example, while there is no evidence of an established or informal quota system in place for women, there is an infrastructural cap to the number of women who can be recruited into the force. Moreover, contrary to popular sentiment, women's interest in policing is not a new phenomenon, as contemporary

Table 5.2 Application Levels and Examination Performance by Sex (2000–2004)

Subject	2000	2001	2002	2003	2004
Applicants	619	369	247	309	162
Male Applicants	311	172	122	145	60
Female Applicants	308	197	125	164	102
Male Sitting Exams	168	120	93	142	40
Female Sitting Exams	113	72	94	136	67
Male Passing Exams	120 (71%)	75 (62.5%)	55	88 (61.9%)	22 (55.5%)
Male Failing Exam	48	45	38	54	33
Female Passing Exams	85 (75%)	56 (77.7%)	70	97 (71.3%)	18 (26.8%)
Female Failing Exams	35	16	24	39	34

Table compiled from data provided by the Human Resources Department of RBPF.

narratives would suggest. In 1950 when the first cohort of women joined the force, four pioneering women were chosen out of 248 applicants. One self-protecting mechanism has always been to cap the number of female entrants.[3] There are fewer facilities for women at the training camp, as well as the station, nationally. The increasing numbers of women within the RBPF, therefore, placed the organizational culture in a state of panic and self-protection.

The present preoccupation with the number of female applicants, on the one hand, suggests a male sense of entitlement to the RBPF, while on the other hand, it does little to help us understand the declining numbers of male applicants. Rather than succumb to the difference distraction, it might be more useful to place recruitment concerns in the context of local gender ideologies and the effect that such ideologies hold for male aspirations, entitlement, and performance. For example, in direct contradiction to questions about women's performance, female recruits have consistently had greater levels of success at entrance exams than male recruits, with the exception of 2004. However, male underperformance should be placed in the larger societal context of males' general academic underperformance in the areas tested within entry requirements (e.g., English, Arithmetic, and Civics; Bailey 2002, 172; Whiteley 2002, 186–187).

Moreover, scholars have argued for the central role that risk taking plays in the performance of masculinity (Chevannes 2001, 161–169). Academic discussions regarding male risk taking behavior (e.g., illegal activity and unprotected sexual contact) highlight such behavior as a significant component of performing masculinity. Unfortunately, they rarely make the much needed analytical connections between these socio-cultural performances of masculinity and male achievement. Interestingly, the RBPF's stipulation that all applicants secure a background check and pass the requisite substance and psychological tests brings these two competing aspects of masculinity to a head. In other words, the socio-cultural performance of masculinity can, at times, be antithetical to the demands of policing.

Furthermore, as policing loses the prestige of yesteryear, it may find itself unable to compete within the wage market. As one male respondent noted:

> There's another thing. (First) there's a salary, (then) you have to pass seven tests[4] .... To be a civil servant for the same pay, you don't have to pass any tests. So, why would you be a police, you can stay home every night, you don't have to face danger? Because the men don't see the salary as so attractive. I've spoken to fellas on the block, and they say drugs will pay them what it gets a police to make in a month. (interview with Police Inspector July 19, 2005)

In the past, remuneration, and occasionally in lieu of adequate remuneration, the cultural capital of policing added to a sense of male identity and place within the community. If policing no longer appears to pay men what they believe themselves to be worth in the labor market and the ties between policing, masculinity, and prestige weaken, we might indeed find fewer men applying to the profession. For those women who do gain entrance, they are constantly called upon to justify their roles and capacity within the RBPF in ways that are not required of their male peers. They are subject to harsh verbal rendition of infractions and the need to work harder to gain recognition.

EMBODIED INTRUSIONS: REPRODUCTION AS LOSS OF MANPOWER

Nothing makes women in atypical professions as hypervisible as the presence of their bodies. As the dual play of hypervisibility and culpability come to rest on women's bodies, women are once again vulnerable to narratives of underperformance and underachievement within the RBPF. In such a constellation, women's embodied difference may serve to limit their possibilities even when accommodations have been made for such difference. As legal scholar Catherine MacKinnon rightly puts it, "How (do) women get access to everything they have been excluded from, while also valuing everything that women are or have been allowed to become?" (1991, 237). When applied to an historically masculinized para-military organization, MacKinnon's sentiments prompt us to think about the ways in which the practices, rules, regulations, and informal procedures of these organizations have been structured over time with the clear awareness that the modal functionary is not a primary childbearer or caregiver.

Without prescribing motherhood for women, it is worth noting that of the women who entered the RBPF in 1950, all retired childless. Alissa Trotz, in her examination of the private security industry in Guyana, observed:

> If motherhood is essentialized, the conditions under which it is carried out must nevertheless be kept invisible, separate from the workplace and subordinate to its demands. (1998, 44)

Here, Trotz captures the societal contradiction of expecting women to mother and to work, but once in the workplace, to mother as unobtrusively as possible. Women's disproportionate responsibility for childcare makes this an impossible tension. The societal devaluation of care work is keenly felt within enforcement agencies in which the valued gender ideology is antithetical to the ethics of care.

The guiding logic of the RBPF ensures that reproduction places a number of hindrances in the pathway of women's advancement within the RBPF. As with all forms of inequity, these hindrances are both formal and informal and infuse the everyday procedures of the organization. These hindrances include extended placement on light duties; the absence of a maternity uniform, which limits women's functional service to the force, and when not in uniform, further marks them as "other" to policing; the lack of childcare facilities to enable night shift and emergency duties; and poor oversight on the physical fitness requirements *for all officers.*

Female officers are placed on light duty as soon as they indicate that they are pregnant (often in the first trimester) and upon re-entry to the service. This practice produces considerable aggravation among supervising officers:

> You see, I had a lady who came to me this morning. She is the driver, she is pregnant. Now, she is not really showing yet, but she comes to me. Now, I have to find out how advanced she is. And if somebody tells you she is pregnant, I have to take her off main policing, and I have to find somewhere to put her. (interview with Senior Superintendent July 12, 2005)

The above response encapsulates the concerns that pervade the RBPF regarding police women's reproductive health and well-being. At the heart of these concerns is the deeply important question of how we should treat women's difference in the workplace. In other words, what impact does women's "potential difference" (childbearing) have on the workplace, and how should women be treated as a result of this potential difference? I use the term "potential difference" because not all are able to perform this difference. Neither do all women, despite capacity, choose to perform this difference.

Members of the Senior Command Team as well as Station/Shift sergeants repeatedly discussed pregnancy as a loss of "manpower" and as an avoidance technique by women to leave the operational aspect of policing:

> Well, it is light duty. If you were to check, you may very well find that if you were to go to the women in headquarters, if you were to check,

they're either pregnant, or they can't work, or they can't work in the sun, or they can't work at night. (interview with member of Senior Command Team July 12, 2005)

In this excerpt, pregnancy is placed on the same sphere as other techniques that women supposedly use to be placed on light duty. Marking pregnancy as a strategy to access light duty simultaneously paints pregnancy as a threat that can potentially unravel the order of things within police enforcement. Scathing but succinct, one senior officer summed the dilemma up in the following way: "Light duty is confusing the force. A lot of women are going that way so that they don't (have to) do this, don't do that . . ." (interview with Senior Superintendent July 12, 2005).

It should not be assumed that pregnancy is a matter that women take lightly in relation to the job, as is evident in the following excerpt:

> I was well into the service, ten years, before I got pregnant. I waited to see how I would handle work and child rearing, getting someone to keep her and so on. Other experiences of other women police and their difficulties. I look at all that before I decide.
>
> *Interviewer:* What kind of difficulties are you referring to?
>
> Like, difficulties of having to run to the nursery, not having the person to keep the child, and I can't work shift. And how to get around the problems of child rearing . . . You know, you will get up in the morning and get ready to go to work and then have to deal with a problem that you didn't plan for. I also had my mother, an extended family, my grandmother. That's why I only have one. You gotta be able to have some persons to help you carry that child or they get away from you. I always tell myself I not going to juvenile court, probation office, I ehn (not) running about with them things. (interview with female sergeant July 18, 2005)

When women opt out into a clerical arena, they in essence attempt to resonate their disproportionate responsibility for childcare with the demands and rigors of the job. The extended family network proves invaluable in facilitating women's entry to this non-traditional profession. While more readily apparent for female officers, male officers with working spouses acknowledge also the value of mothers and grandmothers in providing childcare support.

Since the 1970s, the Caribbean region has provided strong pro-natalist legislation. Barbados provided Maternity Leave under the Employment of Women (Maternity Leave) Act 1976. This legislation entitled employees to not less than twelve weeks after a year of tenure. Additionally, it entitled an employee who, after her leave, becomes ill as a result of her pregnancy up to an additional six weeks, as recommended by a medical practitioner. With stipulated exceptions, the employee cannot be terminated or compelled to resign during this period.

It is helpful to place this concern within the demographic context of Barbadian society. The total population of women within the RBPF is approximately 180. The average age of women who enlisted into the RBPF over the past five years is 23.4 for police constables and 28.3 for special constables. Drawing on wider demographics, the average age of police women within the RBPF falls squarely within Barbados' average childbearing range, which is twenty to twenty-eight. Thus, we can assume that reproduction will matter most for newly graduated police constable recruits within the first five years of their entry into the force (i.e., between 23.4 and 28).

Drawing on demographic trends, too, Barbados' fertility rates declined by 27.9 percent between 1978 and 1999; the present fertility rate is 1.5, and the present population growth is less than 1 percent per annum (Yeboah 2001). It is unlikely, in light of these demographic trends, that these women will, in the main, have more than two children as the literature suggests is the case for Barbadian women who are both educated and employed (ibid.).

These figures, however, will do little to sway the present organizational inclination to assess pregnancy in terms of subterfuge or wastage. Concerns about wastage (i.e. loss of personnel) become even more pressing as regional bureaucracies are pressed to function more "efficiently" under the rubric of Public Sector Reform.[5] Women's bodies, therefore, become the ultimate mark of inefficient productivity. As a result, their reproductive capacity becomes code for a more complex conversation about "time"—who has legitimate access to it and who abuses it and, by extension, the organization's capacity to meet its mission. Marisa Silvestri, in her comprehensive examination of gender and leadership, observes of her own respondents:

> A recurring theme of women's narratives of their career was the significance of "time" as a key resource for achieving a successful career . . . The work of policing, and by implication the lives of police officers, is structured to accommodate a male chronology of continuous and uninterrupted employment. (2003, 90)

Senior male officers described pregnancy as a hindrance to women's capacity to be more "operationally robust." One lamented that women when pregnant are "taken out of the system for six months and if you're not a person who is keen and focus on a career, you can have three four children." Pregnancy not only disrupts law enforcement's idealized chronology but becomes antithetical to a careerist mindset.

We would be mistaken to think that it is a chronology that does not also cost men, admittedly to a lesser degree. Male officers who made it to the top, when asked whether their career trajectory had domestic costs, acknowledged the toll that advancement exerted:

I want to say to you that, being a police officer, I would have been away from my family at critical times. Not necessarily now at this rank, but being an officer has impacted on family life. You will probably get this answer from the older ones as opposed to the younger ones who report sick. (interview with member of Senior Command Team July 12, 2005)

Yet, a closer look at the rhetoric around "time" and loss of "manpower" also reveals a gendered fault line. If we balance the evident anxiety about women's maternity leave with a comparative discussion on other forms of general wastage within the RBPF, there is ample evidence of male officers who exit the system without loss of all benefits and gratuities. A comparison of women's maternity leave against men's general leave might lead one to argue that while women *need* time, men *get* time. In this regard, it is instructive to look at the gender differentials both statistically and in following the narrative regarding wastage within the RBPF.

A Station Sergeant within the RBPF succinctly assessed the impact of male wastage in the following discussion:

Pregnancy has an impact, but it is their right and it is something that we must be prepared for, because at the same time, we have officers who are sick and they ain't pregnant. Pregnancy is something that you will be back at work, but the guy who has been sick for six months, you don't know when he's coming back. So, you can't look at pregnancy as the only factor that will cause you to be short of manpower. Yes, we have women who are pregnant, but we also have men who are sick and stay out longer than a particular time. It is their right, and the reason we haven't had to deal with it before is because men don't get pregnant. People may not be on extended sick leave, but they use sick leave at critical times. We had a girl who was pregnant and sick in the morning, and we flexed her time to

Table 5.3 Male Wastage on Medical Grounds

Year	Total Wastage	On Medical Grounds	Percentage of Total Male Wastage
2000	81	35	43.2
2001	65	23	35.3
2002	77	23	29.8
2003	53	17	32.0
2004	47	19	40.4

accommodate her. And 90 percent of the time she still comes. *We have that approach recognizing that she is one of us.* (emphasis mine; interview with Station Sergeant July 15, 2005)

Total wastage incorporates categories such as age retirement, death, secondment, and transfers—all categories which have been recognized as a legitimate rite of passage or move toward advancement. However, leave on "medical grounds" accounts at a minimum for 30 percent of the men who are unavailable for duty in any given year. Despite the fact that men, proportionally and in absolute terms, account for the greater degree of wastage, the excerpt above was the only mention of men's infraction of time and contribution to overall wastage in the entire interview process.

"YOU CAN'T PLACE YOURSELF: SOMEBODY GOTTA PLACE YOU": WOMEN AND ORGANIZATIONAL SEGREGATION

I want to revisit my earlier reference to MacKinnon's finely tuned articulation of the importance of providing women with access to previously closed spaces without denying their difference. Throughout the interview process, senior administrative officers repeatedly rendered pregnancy as an act of willful subterfuge to access light duty. There was rarely any organizational reflexivity regarding the organization's own procedural culpability in facilitating women's access to light duty generally. There was, though, a clear organizational awareness that light duty (clerical work) was not exactly a means of fast-tracking one's way through the organization.

Women's access to light duty in the midst of administrative blustering recreates a form of organizational segregation in which women become token officers. This encroaching segregation begins with a strident belief in the "objectivity" of the criteria for advancement and ends with the willful ignorance of the ways in which assessment processes are already gendered and, further, that these gendered processes have been normalized. The normalization of these gendered processes masks the preference for masculinized practices and areas of operation as the means for advancement and, as I will argue here, masks the organization's complicity in allowing women to opt out of these areas of operation. This dissonance and tension between the integrity of procedure and the outcome for women is evident, for example below, in the narrative progression of a senior officer when asked about promotional criteria:

> I would prefer to say that you are performing satisfactorily in your area of assignment. If I'm assigned to CID (Central Intelligence Division), then my supervisor should expect me to go out there, deal with matters of crime, solve a few cases. If assigned to clerical work, my supervisor should expect

me to perform in a way that is acceptable. So, I don't want to say only hard policing because it depends on where you're working . . . (interview with member of Senior Command Team July 12, 2005)

Yet, later in the interview, the officer noted:

> If you're working as a clerk and another person is working major crime and it came down to one vacancy, then perhaps the person working crime will get the tip. But I don't want to say that because individuals do not have control over where they are transferred. So judge me according to where you've placed me. (ibid.)

Yet, such sentiment, despite its appearance of neutrality, is deeply gendered as a result of the occupational segregation that obtains within the RBPF. As evident in the first excerpt, the perception of women's underperformance by the leadership and some rank and file within the RBPF is governed by an argument that suggests that women "gravitate" towards clerical work. What appears on the surface as an argument of women's "choosing" clerical (i.e., non-operational) duties is, though, a much more complicated discussion that implicates the RBPF's gendered organizational culture through aspects such as gender stereotyping in task allocation and management practices, which do not efficiently detect when officers have not maintained necessary ongoing training programs (e.g., firearms training).

There is a great concern within the RBPF that women gravitate toward and/or choose to do clerical work rather than participate in the more operational aspects of policing. The reasons given for this occurrence are numerous: that women are afraid to interact with the public due to possible volatility; that they are better suited for office work; and, as I have already discussed, that women gravitate toward this arena due to pregnancy. Is women's "choice" or "gravitating toward" feminized tasks in the workplace a neutral decision? Is it that women choose clerical work as a result of RBPF's gendered culture that establishes the parameters of women's legitimate spheres of activities? In other words, why is it that women are concentrated within clerical or light duty within the RBPF? Is the dominant sense that "woman doan wan work" an adequate response to these questions?

Legal theorist Vicki Schultz argues that sex segregation in the workplace (e.g., gravitating toward clerical work) is not individually determined. Schultz maintains that merely assuming that women "lack interest" simultaneously suggests that the organization is not complicit in encouraging this "lack of interest" (1990, 1753). She goes on to argue that "Society has long viewed . . . women as inauthentic workers, uncommitted to wage work as an important life interest, and a source of identity. This view has justified relegating them to dead-end, female dominated jobs at the lowest rung of the economic ladder" (ibid., 1756). Women's "individual interest/ choice" argument successfully diverts attention away from the ways in which employers choose to not disrupt patterns of gender segregation in the workplace. That is, we must examine the ways in which employers organize their workplaces that disable women from forming an interest in non-traditional work (ibid., 1842).

One of the ways that this gendered structuring occurs within the RBPF is via a heavy dependence on pre-existing gendered patterns of subject allocation and streaming at work within high schools. A large body of research shows evidence of sexist patterns within curricula, as well as subject allocation within the Anglophone Caribbean region (Parry 1996; Leo-Rhynie 1997; Bailey 1997; Bailey 1998). Female police officers are very aware of the fact that they will be more quickly placed to perform office duties than their male counterparts, as reflected in the following response from a female police officer when asked about female officers' "preference" for clerical duties:

> Well, you have two sides to that question. You gotta remember that in the force you can't place yourself, somebody gotta place you. How many men will you find with clerical education because it is a whole education talking about . . . you talking about typing, office procedures. Come on. How many men would you . . . go into a class at school and tell me what you see. (interview with female police constable July, 20 2005)

Women do not *choose* these arenas because of pre-existing patterns of sexism within the secondary system, but rather, as this officer astutely encourages us to observe, the organization manipulates these pre-existing patterns and perpetuates a gender-segregated environment within the workplace. That women came into the force with computer skills lent a tone of "fair play" to essentialist arguments that women came with a predisposition for clerical work. The following provides an example of the deleterious effects of this naturalized perspective on women:

> I think they try to get into the offices, and it's true that more of them are trained in computers than males, and this is how they could get away. Sometimes, you want this computer work done, but everybody trying to getting into the office or not work at night. When you interview them for the force, they tell you they will do all sorts of things. But when they get in, they don't want to do anything. (interview with member of Senior Command Team July 12, 2005)

With the valued discourse being that of operational policing, the reliance on existing sexist patterns of high school curricula formulation as a rationale for placement within the force contributes significantly to the systemic marginalization of women within the RBPF.

What seems to be willfully ignored when rendering women as agents of choice in non-operational duties is the issue of "placement": "You gotta remember that in the force, you can't place yourself, somebody gotta place you;" "Sometimes you want the computer work done and that's how they get away"; and "That's where we put them because we still do a lot of office work." The RBPF as a para-military organization is governed by clear delineations of the lines of authority. Nonetheless, in the midst of these clear lines of authority, there is still the dominant narrative that (1) women choose clerical work, or as we will examine later in this chapter, that (2) subordinate females seduce their superiors for favors and preferential treatment in task allocation (i.e., removal from the operational side of functioning).

Some female officers do not wish to remain as clerical staff, just as not all male officers are interested in the operational side of policing. Yet, because of the gendered environment created by the overdetermined allocation of women to clerical duties in the RBPF, women more readily use this sphere (e.g., by stressing medical need) to strategically exit from the operational side of policing. In other words, this is an example of women using the gender-segregated bias constructed within the RBPF to their own advantage. This practice of negotiation in the workplace operates with a great deal of functional similarity to more masculinized strategic practices, such as networking with mates from one's cohort at police academy (i.e., one's batch) and "liming" (informal social time) in male spaces after work (e.g., bars, cricket clubs, and football).

However, for women with long-term promotional aspirations, the gendered segregation of women into clerical duties does not favorably position them to achieve the same aspirations.

Despite the argument that women now have more time to study for the exams once in the promotional zone, without the operational element to demonstrate knowledge of the job and, by extension, the ability to instruct, supervise, and advise subordinates, women who do not function within the operational side are unable to vie successfully. This

Table 5.4 Female Officers' Promotion in the RBPF (2000–2004)

Year	No. of Women Promoted/Rank	Female Strength
2000	2 (SGT)	106
2001	2 (SGT)	124
2002	6 (SGT/INSP)	153
2003	3 (ASP/STSGT/SGT)	176
2004	4 (ASP/SGT/STSGT)	184

resonated throughout the interview process, as the following response indicates:

> You will find that it is difficult to move from constable to sergeant without doing hard policing. But when you look at inspector above, what they are looking for are management skills. (interview with member of Senior Command Team July 17, 2005)

So, while the employer cannot say that women do not apply, the employer can, via constructed segregation, say that women, if confined to clerical work, are not best suited. Hence, we can conclude that neither did larger numbers of women within the RBPF translate into immediate rewards for women, nor did these numbers better position women for success. Rather, the increasing numbers of women placed the institutional culture into a state of panic and self-protection and, as such, served to place women in the RBPF in a hostile environment in which they are constantly called upon to justify their roles and capacity, subjected to harsh verbal rendition of infractions, and required to work harder to gain recognition.

FLEXING POWER: SEXUAL HARASSMENT AND THE MAINSTREAM

In the preceding discussion, I've attempted to show the ways informal and formal policies, interpersonal dynamics, architecture, and institutional narratives have all worked to construct and frame women's bodies as a problematic kind of difference within the workplace. I've also highlighted the ways in which the framing of women as different holds significant material consequences for the extent to which gender equity is actually achieved within the "mainstreamed" workplace (e.g., job mobility, training, and placement). I now want to turn our attention to an examination of the ways in which sexual harassment functions as a very specific kind of containment of women's embodied difference in the workplace.

That we are able to name a set of workplace practices as sexual harassment, and further, that we can identify these practices as harmful, is due in no small part to the work of feminist legal scholar Catherine MacKinnon. In her groundbreaking text *Sexual Harassment of Working Women: A Case of Sex Discrimination* (1979), MacKinnon identified sexual harassment as

> ... the unwanted imposition of sexual requirements in the context of a relationship of unequal power. Central to the concept is the use of power derived from one social sphere to lever benefits or impose deprivations in another. (1)

MacKinnon's early work offered a number of insights that today remain foundational to feminist conceptualizations of sexual harassment. While sexual harassment may be experienced at the individual level, MacKinnon insisted that sexual harassment not be individualized (1979, 83–90). In other words, sexual harassment was an expression of a gender hierarchy to which women as a collective group were "systemically vulnerable" (1979, 4). MacKinnon saw the harm of sexual harassment as a form of sex discrimination at two interrelated levels: first, as a sexual imposition on women and/or sexualization of women in the workplace and, second, as an imposition that placed severe limits on women's possibilities in the workplace (ibid.). For MacKinnon, sexual harassment's harm was but one example of a number of social manifestations that perpetuated men's control of women's sexuality. In MacKinnon's work, we find an engagement with the multiple deployments of sex for the purpose of subordination. This included sex as activity as well as an inclusion of discriminatory practices against individuals who are marginally positioned within the sex/gender dichotomy as well as the use of sexual imagery to structure a hostile institutional culture. This range of engagement has prompted legal scholar Katherine Franke to call MacKinnon's work an "anti-subordination" approach (1997, 726).

The preponderance of male-on-female examples of sexual harassment, coupled with MacKinnon's insistence that sexual harassment reflected a gender hierarchy that privileged males, opened her work to the criticism that women were not only particularly vulnerable but also that women were the only victims of sexual harassment. As Franke observes, a less-than-thorough understanding of an "anti-subordination" approach could lead one to conclude that "Men, therefore, cannot be the victim of sexual harassment because their sexual objectification, either at the hands of women or other men, would not mirror this subordinating social hierarchy" (1997, 728). Women's advances in the workplace may prompt the contemporary reader to question MacKinnon's, at times, rigid binary of women's inferior positioning to men's hold on power, not an illegitimate impulse. Yet, MacKinnon vehemently opposes the idea that sexual harassment is a "natural" expression of human (read male) sexuality, arguing that such a sentiment, whether encoded as a legal or populist expression, demands that we confront "an inequality in social power (that is) rationalized as biological difference" (MacKinnon 1991, 92).

In addition, Franke identifies two additional, though limited, rationales for sexual harassment as an act of sex discrimination. The first, "but for sex," acknowledges that harassment would not have occurred but for the person's sex (i.e., were she not female, the offense would be less probable). The second, "because of sex," identifies an act as sanctionable because the offending behaviors are of a sexual nature (i.e., causing the sexualization of, or sexual imposing on, an individual in the workplace). While these three arguments have individually informed successful petitions, Franke questions the extent to which they accurately account for the wrong of sexual harassment (1997, 693). Franke does not contest the discriminatory force of sexual imposition of men on women because of their sex. Neither is her aim to dismiss the material inequity that women experience as result of sexual harassment, nor is she blind to the fact that some subjects are more vulnerable to acts of sexual harassment than others. Rather, Franke maintains that despite the practice or rationale, the common intent of each of these instances of harassment is but one simple, dangerous desire: gender conformity. Sexual harassment, Franke notes, is

> . . . a technology of sexism. It is a disciplinary practice that inscribes, enforces, and polices the identities of both harasser and victim according to a system of gender norms that envisions women as feminine, (hetero)sexual objects and men as masculine, (hetero)sexual subjects. This dynamic is both performative and reflexive in nature. Performative in the sense that the conduct produces a particular identity in the participants, and reflexive in that both the harasser and the victim are affected by the conduct. (ibid., 694)

Franke's account perceptively identifies what is at stake when we refuse to name sexual harassment as an act of discrimination within the workplace, namely, that the workplace becomes a site of gender terror where variegated codes of sex and sexuality are deployed to secure hegemonic (as distinct from normative) performances of sanctioned gender identities.[6] Franke's call for a shift from sexual harassment to gender harassment immediately expands the permutations and contexts of offenses that can be accounted for as discriminatory.

Without sacrificing women's peculiar vulnerability to sexual harassment, "gender harassment" extends this boundary to simultaneously account for similar impositions among different sexual permutations (e.g., same-sex encounters and homosexual/heterosexual encounters; ibid., 694). Within a gender harassment framework, the sexual aggression of sexual harassment can result "because of sex" (as activity) but need not be confined to sexual physical encounters, as is the case, for example, where sexual aggression can terrorize feminized masculinities into a particular performance of hegemonic masculinity. Similarly, while harassment acknowledges that some individuals would not experience sexual harassment but for their sex, Franke's use of gender harassment, by allowing for the ways different gender identities might be potentially squelched, does not confine the discussion to a heterosexual binary. These conceptual explorations are compelling in their recognition of the force that attends gender-based power and for their expansion of bodies and contexts that are differently vulnerable to harassment in the workplace. Franke's discussion of "gender harassment" adds a conceptual depth to "sexual harassment" foundational to this chapter's subsequent discussion.

SUBVERTING THE CENTER: SEXUAL HARASSMENT LEGISLATION AND THE ANGLOPHONE CARIBBEAN

After at least a decade of feminist examination, criticism grew that gender mainstreaming became too amenable to what Bacchi has called a "compensatory agenda" (2001, 19). Labeling gender mainstreaming as compensatory identifies its inability to realize its radically transformative promise, defaulting, rather, to the technical propensity of implementing measures that ameliorate without fundamentally restructuring the systemic factors that have produced and perpetuate existing inequity (ibid.). In Chapter 3, drawing on a case study of the Trinidad and Tobago Gender Affairs Division, I suggested that, contextually, such transformative possibilities have been hindered by a number of hostile bureaucratic manipulations. These practices, I suggested, are anything but the ideal type of Weberian model, governed as they are by partisanship, political expediency, and willful obstructionism of mainstreaming efforts. I also argued that these practices are not without individual and institutional ramifications: psychological incarceration and physical exhaustion, on the one hand, and organizational hostility, on the other hand.

There are a number of processes to which we must attend if we find value in gender mainstreaming's stated goal that all sectors should work to ensure that women and men are equally incorporated, and, I would add, able to achieve commensurate outcomes. These should not only include getting the numbers "right" as we account for diversity and statistical change but should equally incorporate procedures and messages that reorient the mainstream's quotidian thinking on which bodies are entitled and allowed to move unencumbered throughout organizational spaces. For, with the precision of a cartographer's pen, sexual harassment uses sex to demarcate place. In other words, sex (both as activity and gender) becomes an encumbrance on the spaces that "alien" bodies may occupy. Therefore, while gender mainstreaming emphasized inclusion as desirable, I am making a strong argument here that additional instruments are required to ensure the well-being and expand the possibilities of those marginalized bodies once they are "included." Gender equity in the workplace raises the stakes on senior administrators. Equity demands that we give sustained evaluative attention to the multiple organizational principles and assumptions that perpetuate hierarchies of difference within the workplace. It also requires the development of specific corrective and reorienting mechanisms for those who are mainstreamed.

Here the work of the feminist organization Coalition Against All Forms of Sexual Harassment (CASH), which began in Barbados in April of 2003, has been pioneering in its mandate and structure. CASH functions as the first feminist organization in the region established with the sole purpose of securing legislative reform in the area of sexual harassment. An advocacy group, its finite mandate was to ensure that Barbados' Draft Sexual Harassment Bill address the harm of harassment in as comprehensive a manner as possible. This included a call for conceptual expansion, delineation of responsibilities for harassment-free environments, and appropriate protections and remedies. To this end, the organization committed to using its multi-disciplinary expertise to facilitate consultations and data gathering, to critique and draft comprehensive sexual harassment legislation, and to mount and disseminate a series of public awareness and educational materials.[7]

While Barbados has yet to secure sexual harassment legislation, between 1991 and 2000 four territories in the Anglophone Caribbean initiated some form of sexual harassment legislation (See Table 5.5).[8] Varied in their degree of comprehensiveness, such legislative interventions range from comprehensive sexual harassment legislation in the case of Belize to statutes within anti-discrimination or equal opportunity legislation in the cases of Guyana and Trinidad and Tobago. Of existing national legislation, only Belize and Guyana address a "hostile working environment" as a feature of sexual harassment. Beyond these exceptions, sexual harassment has been generally understood as *quid pro quo* sexual harassment. In the case of Trinidad and Tobago, sexual harassment is not explicitly addressed; however, the equal opportunity legislation bars any discrimination "in the terms or conditions on which employment is offered" and continues with an exploration of the parameters and exceptions to possible discrimination "based on sex."[9]

With the exception of the Bahamas, which designates sexual harassment a criminal offence, the trend has been to redress sexual harassment through civil remedies (Robinson et al. 2003). While the Bahamas uniquely requires the approval of the Attorney General before a case may proceed before the court, where treated as a civil matter, a variety of adjudicating and conciliation mechanisms exist.[10] The present proclivity for conciliation mechanisms and the preference for recognizing *quid pro quo* sexual harassment to the exclusion of a hostile condition of work approach warrant some attention. Taken in tandem, these two commonalities arguably encourage us to see the harm of sexual harassment as residing only in instances of assault, violation, or threats of such as a contingent feature of employment or reward. Yet, when these violations occur, conciliation becomes the first or singular response. It is difficult to not see such a juxtaposition of violence/mediation as anything but a suggestion that the violation, assault, or imputed gendered disadvantage was at least a "misunderstanding" that can be resolved with effective mediation.[11]

A number of gendered assumptions are at work in this juxtaposition of violence/mediation: that the state carries the ultimate "objective" authority to determine the legitimacy of sexual harassment; that *prima facie* evidence notwithstanding, it is not a prioritized use of the court's time; and that the violence of sexual harassment lies primarily in the trauma of assault, violation, or threat rather than everyday hostility where sex, sexuality, and gender are all variably manipulated to prevent access to those who have been gendered as minorities in the workplace. If these assumptions are true,

Table 5.5 Overview of Sexual Harassment Legislation in the Anglophone Caribbean

Country/Model	Definition of Sexual Harassment	Extent of Coverage	Liability	Punishment/Remedy
CARICOM Model Legislation on Sexual Harassment	Unwelcome advances, requests, suggestions, innuendos in exchange for employment, rewards; no discussion of hostile environment	Employment Education Accommodation	Employer responsible for harassment-free environment	Guilty of an offence punishable by fine not exceeding $5,000 (act of victimization)
The Bahamas Sexual Offences and Domestic Violence Act (1991) Part 1: Sect. 26	Solicitation of sexual favors in exchange for rewards, admissions, employment, advantage; no discussion of hostile environment	Employment	Individual who has contravened Act whether in position of employee or employer	$5,000 and/or imprisonment for two years Prosecution cannot proceed without Attorney General's approval
The Belize Protection Against Harassment Act (1996) Rev. 2000	Unwelcome sexual advances Employment contingent on compliance Hostile offensive environment Unsolicited gestures, comments Display of offensive materials not necessary for institutional purposes	Employment Institutions (e.g., students, wards, and inmates) Accommodation	Employer responsible for harassment-free environment Liable once informed Individual who has contravened Act	Many forms of redress, including but not limited to: • ordered to desist • designated binding-over period in which conduct must not occur; if it does, individual treated as in contempt of court • compensatory damage adjudicated in appropriate jurisdiction as civil debt • fine and/or imprisonment for false claims Court not empowered to reinstate terminated complainant

(continued)

Table 5.5 (continued)

Country/ Model	Definition of Sexual Harassment	Extent of Coverage	Liability	Punishment/ Remedy
The Guyana Prevent of Discrimination Act (1997). Part 3: Sect. 5–8	Creation of a hostile environment. Unwanted conduct of a sexual nature as condition of employment	Public and private sector workplace Provision of goods and services	Individual who has contravened Act (employer, co-worker) Supplemental remedies from employer for indirect or direct losses result from contravention	Unless penalty otherwise provided, liable to fine not exceeding $20,000 GD Successful complainant may apply for damage, inclusive of reinstatement if employer and aggrieved agree
The St. Lucia Equality of Opportunity and Treatment in Employment and Occupation (2000)	Unwanted conduct of a sexual nature in the workplace or in connection with the performance of work which is threatened or imposed as a condition of employment on the employee or which creates a hostile working environment for the employee	Employment	Employer, managerial employee, or co-worker	Successful complainant may be reinstated by court, whether vacancy filled or not Damages from employer
The Trinidad and Tobago Equal Opportunity Act (2000)	Not explicitly addressed 11(1) no discrimination on the grounds of sex (List of exceptions e.g., purposes of authenticity in entertainment)	Employment (inclusive of goods and services) Criterion also established for educational establishments (exceptions outlined for single sex schools)	Any party who has contravened actionable aspects of Act	Conciliatory approach If not amenable or conciliation fails, Tribunal initiated Tribunal invested with power to award compensation, fines, damages

Table compiled from respective country legislation and "Sexual Harassment and the Law in Barbados," 2003, prepared for the Coalition against All Forms of Sexual Harassment by T. Robinson, K. Clarke, and N. Walker.
[1] All monetary penalties rendered in local currency.

then they reflect cultural constraints that exist in the wider society and remind us that while law does have the capacity to adjudicate culture, it is simultaneously a cultural artifact. These gendered assumptions matter to women and other marginalized subjectivities who may be "mainstreamed" into atypical professions and sectors.

Examining the Jamaican organizational context, social scientist Jimmy Tindigarukayo found that the absence of sexual harassment legislation contributed to a general disregard of the need for organizational protections against harassment. Tindigarukayo observed that the absence of national legislation facilitated a general disinclination to implement sexual harassment policies at the organizational level, as well as under-reporting by victims since, without legislation, victims do not expect any action to be taken against perpetrators (2006, 103). Tindigarukayo noted that reports of sexual harassment were more likely to occur when someone's job was threatened, suggesting, as I explore later, that victims of sexual harassment find strategies of circumnavigation as a coping mechanism for the hostilities of the work environment. Tindigarukayo's study calls for legislation that acknowledges that men are also victims of sexual harassment. However, as I will discuss shortly, the suggestion that men are victims of sexual harassment may sometimes function as a rhetorical strategy to deflect attention from addressing sexual harassment as a problem—that is to say, it is not really such a problem because women are harassers or women become harassers through practices of lesbian baiting.

Focusing on gender equity in the RBPF, I explore the peculiarities of integrating gender minorities into historically masculinized professions such as policing. So, while sexual harassment debates predate formal mainstreaming discussions, I argue that mainstreaming produces a certain vulnerability to harassment, and there have been no preventative or corrective mechanisms for such a vulnerability within the domains where mainstreaming discussions occur. The nature of power that is troubled by mainstreaming agendas is such that these vulnerabilities fall disproportionately on the bodies of minority subjectivities; as such, I see the need to connect these two domains as particularly urgent. If mainstreaming is to go beyond numerical incorporation, then policy decisions must be explored in tandem with a wider array of legal instruments, if only because the repercussions of effective mainstreaming sometimes exceed what can be addressed by policy and planning.

AMBIVALENCE IS NOT AN OPTION: SEXUAL HARASSMENT IN ATYPICAL PROFESSIONS

Sexual harassment is about demarcation. It is a memo that is sent to remind the victim of her/his place, to enforce the script for "appropriate" gender performance, to plot with precision the topography of institutional power. These censors are more likely to be unleashed as women and other gender minorities engage in occupations that have been masculinized and male dominated (Gutek and Morasch 1982), where sexist attitudes exist within an organization (Fiske and Glick 1995), and where practices and strong beliefs about gender-based occupational stereotypes and sex-matching prevail (Glick 1991).

Psychologists Susan Fiske and Peter Glick explored the dynamics of sex- based discrimination through their formulation of an Ambivalent Sexism Inventory (1996). There are two defining features to Fiske and Glick's Inventory, still very much in circulation since their early exploration in 1994: first, that sexual harassment is differentiated; and second, that the harasser's motivations matter in helping us distinguish one form of harassment from the other. The use of "ambivalence" is the authors' attempt to explore the ways in which men vacillate between the desire to dominate or be intimate with women. As women enter the workplace, Fiske and Glick maintain, men begin to experience similar forms of ambivalence toward the women with whom they come into contact (1996, 96). When this ambivalence manifests through sexual harassment, the authors argue it is important to distinguish between different permutations of cognitive motivations since these motivations result in very different practices of sexual harassment. The two major categories in their Inventory are hostile and benevolent sexism, which they define as follows:

> Hostile sexism refers to sexist antipathy toward women based on an ideology of male dominance, male superiority, and a hostile form of sexuality (in which women are treated merely as sexual objects). Benevolent sexism refers to *subjectively positive*, though sexist, attitudes that include protectiveness toward women, positively valenced stereotypes of women (e.g., nurturance), and a desire for heterosexual intimacy. (ibid., 98)

Noting that benign sexism does not occur to the exclusion of hostile sexism, the authors draw on ambivalent sexism to mark the shifting pendulum between men's desire to impress/suppress. I find their work useful in its attempt to engage the motivations of perpetrators of harassment, as the authors themselves note "ambivalent harassment may be particularly insidious because the man can readily justify his actions to himself as not harassment" (ibid., 100). These motivations provide useful insight into the opposition toward sexual harassment policies within masculinized work environments. For example, Fiske and Glick suggest that three common impulses characterize both hostile and benevolent sexual harassment: (1) paternalism, in which men become/act as protectors or disciplinarians toward women; (2) gender differentiation, governed by men's desire to demarcate their masculinity as not merely different from, but superior to women; and (3) heterosexuality, which drives both intimacy and dominance impulses (ibid.). This ambivalence is a movement between hostile sexism, characterized by male domination, and benevolent sexism, which reflects "an earnest (subjectively benevolent) form motivated by a genuine desire for lasting heterosexual intimacy" (ibid., 99).

Yet, Fiske and Glick's Ambivalent Sexism Inventory is severely handicapped in two fundamental ways: it lacks a transformative component and remains rigidly heterosexist. Because sexual harassment works to preserve gender-based power within the workplace, it is not a narrative that can be subversively told from spaces of privilege. So, while cognitive motivations are instructive, the use of "benevolent sexism" erases the trauma of harassment as experienced by the recipient.[12] Sexual harassment becomes benign when clad with narratives of "romance" and intimacy. The primary emphasis on motivation silences women's voices from negotiating the meanings of harassment. Consequently, the juxtaposition of "harassment" and "benevolent" in their work leaves the power differentials of the workplace uncontested.

At times, ambivalent harassment borders on positioning women as the cause of harassment. For example, women who fall into different subtypes (e.g., "sexy" or "traditional") may be more vulnerable to instances of harassment. They note that "women who fit both the 'sexy' and 'nontraditional' categories (e.g., a physically attractive, ambitious female stockbroker) are doubly at risk because *they are the most likely to arouse the volatile mixture of intimacy-seeking and hostile motives that characterizes ambivalent harassment . . .*" (emphasis mine; ibid., 103). The formulation of the sentence positions the recipient of harassment, in this case women, as both provocateur and victim; a curious ambivalence, indeed, in which women rather than men are called on to account for the hostile nature of the workplace.

The Inventory's heterosexism reinstitutes harassment within a gender binary of what men do to women (lesbian or heterosexual) and, therefore, helps little to further an understanding of sexual harassment along a wide array of gender performances and power struggles within the workplace. Admittedly, the authors do advocate for clamping down on hegemonic masculine cultures within the workplace as an organizational remedy. However, the authors' suggestion that we explore intimacy, albeit misguided and subjective, as one of the cognitive motivations for harassment, inscribes a degree of passivity to male heterosexist, hegemonic behavior (e.g., male entitlement to women's attention). Understanding motivations is useful insofar as men and women define experiences of harassment differently. Yet, suggesting that we understand harassment through the innocuous language of "benevolence," "intimacy/romance," and "ambivalence" minimizes the tyranny that the recipients experience, limits the ways in which corrective action can be secured within the workplace, and creates false affinities between "harassment" and "desire."

I am not suggesting that all workplace sexual encounters are harassing ones. To take such a position is equally as paternalistic. It leaves no room for anyone to exercise individual sexual agency in the workplace and presumes that all sexual encounters are insidious in nature. Non-harassing encounters raise questions about the effectiveness of absolute sanctions against sexual activity between co-workers while potentially making it more difficult for organizations to minimize their liability to claims of sexual harassment. Non-harassing interaction offers important research opportunities such as the effects of increased contact among workers (Gutek et al. 1990), romance, and its effects in the workplace (Schaner 1994), or the effects of non-harassing intimacies on productivity or incidences of sexual harassment. However, it is slippery, particularly for those most vulnerable to instances of sexual harassment, to see it as vacillating between benevolence and hostility. Unless there is a willingness to interrogate the implications of privilege, those who hold power in the workplace may find themselves reinforcing discriminatory practices through narratives of benign intentionality ("I didn't mean to harass"), trivialization ("having fun"), or the difference distraction ("We didn't have these problems before").

To legislate that all sex is inappropriate sex confines sexual harassment to an abuse of an act. Franke challenges this limitation through her examination of sexual harassment as an assault on one's gender identity. I think, though, considerations of sex as a violation will be greatly aided by considering the question of dignity and fairness. To legislate against sex in the workplace presumes that all sexual encounters and symbolic references are deployed to rob a person of dignity and to undermine equitable practices. I draw here on Kathrin Zippel's discussion of European legislative trends on sexual harassment. For Zippel, the common characteristic of continental legislation is a focus on "fairness in the workplace," which she argues "has the potential to go beyond the narrow focus on sexual conduct and instead to draw attention to the both exclusionary practices and institutionalized structures in the workplace that perpetuate these practices" (2004, 62). Consequently, sexual harassment's offense is not just that it is sex-based, but that, in addition to coercing gender conformity, it fundamentally attempts to violate one's personal integrity and dignity (ibid.). Zippel's focus on dignity opens room for a less conservative and totalitarian approach to sexual harassment as sexual conduct. In other words, it is not that all sex is bad. Rather, it is the purposeful deployment of sex and sexual practices with the intent to shame, coerce, and marginalize any group that may be designated as a gender minority that harms individuals in the workplace.

SEXUAL HARASSMENT AND THE ROYAL BARBADOS POLICE FORCE

Beyond outlining remedies and protections, sexual harassment legislation serves a number of important social benefits. On the one hand, the passage of legislation serves both as a catalyst for change in organizational policy and practices and as a preventative function. On the other hand, the passage of legislation opens a space for public debate and dialogue. Along with remedies and protections, these two functions are critical to collective reflection and behavior modification.

Addressing sexual harassment in the RBPF presents a catch-22; because there are no channels to facilitate formal reporting procedures that specifically address sexual harassment, it follows that there are no formal reports of sexual harassment. So, the absence of formal evidence is offered to suggest that sexual harassment does not occur. Sexual harassment legislation notwithstanding, there is a dearth of information at both the empirical and conceptual levels regarding sexual harassment in the Anglophone Caribbean. Interviews within the RBPF mirrored some of the findings in extant literature: nebulous conceptualization of sexual harassment, trivialization of harassment, and mistrust of reporting for fear of further victimization. However, the RBPF interviews provided additional insight into the use of lesbian baiting as a strategy of harassment against women who enter nontraditional professions.

Anthropologist Carla Freeman's study of data processing centers in Barbados highlighted the effect that vertical fraternizing had on a processing center's morale. Among the sentiments that her female respondents offered was the prevailing sense of gender as nature (men jus' can't help themselves). Freeman, citing a female trade unionist from the Barbados Workers' Union, captured the murkiness of naming and identifying the parameters of sexual harassment in the workplace. Her respondent framed this difficulty as follows:

> Women don't know exactly how to define it, most think of it only in a physical sense, not the subtle effects and mechanisms, the stress that is caused and the psychological effects of it. The women are afraid to acknowledge their personal experiences outright—that the authority figure makes certain gestures at you—out of a sense that it would be seen as boasting [and the sense that] their co-workers would say, "He's never done that to me. What makes you think you are so sexually attractive that he would do that to you?" They would start to doubt themselves. You know, it's sexual harassment, though. (2000, 183)

In this very rich response, instructive at a number of levels, the respondent identifies women's propensity to confine sexual harassment to instances of *quid pro quo*, extends the definition of harassment to capture less overt expressions, and additionally, begins a critique of any inclination to equate harassment with attraction.

The common association of sexual harassment with sex acts heightens the tendency to see sexual harassment only in terms of exchange. So, while members of the RBPF offered varied understandings of what constituted sexual harassment, male respondents more readily offered descriptions of *quid pro quo* forms of sexual harassment. When asked if they thought that a hostile working environment could be considered a form of sexual harassment, male officers not only dismissed this as an illegitimate form of harassment, but further, observed that to complain about that form of harassment was antithetical to the "rough and ready" nature of the police environment:

> I see sexual harassment as a direct action or process. I never thought of it as passive. In this job, to have something like a calendar affect you, I don't know, considering the things that a person has to face sometimes everyday. For someone to be affected by the images of the calendar doesn't make sense. I hope you don't think I'm being callous. (interview with detective July 17, 2005)

Or again,

> Well, I would want to raise a daughter who is strong and able to get past a situation about somebody telling her about her body because I think that that is going to happen in an organization, whether it is the police force, C.O. Williams, or walking down Broad Street.[13] (ibid.)

These comments draw attention to two impossible constraints against women as officers and as women. The first is the sense that women's resistance is construed as an unwillingness to "play along with the boys," thereby, making them vulnerable to once again being seen as illegitimate police officers. Second, that Barbadian women's objection to harassment generally is an indication that they are out of step with their broader society and Barbadian culture, in other words, aliens or, even worse, feminists.

SEXUAL HARASSMENT COMPLAINTS AND ILLEGITIMACY

To suggest that female officers are ill-prepared for policing when they name sexual harassment as harm serves to make women more unwilling to report for fear of being accused of "child-like" or "girlish" behavior, both being the ultimate markers of un-belonging in a masculinized profession. In these instances, women and other gender minorities by virtue of their non-conformity are perceived as a threat to those informal and formal practices of camaraderie that build community through an enforcement of gender codes.

The irony is, of course, that the actions with which women are being required to be complicit are not actions or gestures that are extended to "one of the boys." Rather, these gestures are extended to some women, to some men—often men designated as effeminate, and to men and women who challenge heterosexuality. That women are at odds with male renditions of sexual harassment is immediately evident by their definitions of sexual harassment. Female officers' understanding of sexual harassment was broader and more varied than men's:

Sometimes, I think it ain't even touching, just the person talking or saying certain things I would take as sexual harassment. Like if someone just come and tell me . . . how I would like to do this to you, I would take that as sexual harassment because you and the person ain't the kind of friend that they should come and tell you that.

Well, to tell you the truth, as a female, when your supervisor is making advances towards you for favors, keep nagging you, touching you.

We have males who approach you on a daily basis at all levels, we have constables, sergeants, inspectors. They approach you and try to get you to go out with them, and they tell you things about parts of your body or tell you what they would like to do with you. Those are some of the things you experience on a daily basis.

Yes, it exists. It may not be so bold, but it is subtle, and you may not see it with your own two eyes. Why is that person in that person's office so often? Why does he have to call her into his office ever so often? Is it just work related, or is he trying to get something else? That's the subtle approach because every time you look around, this person is in the office with him and all of a sudden the person doing different work, office work, and duties change, favors now come in. (interviews with female police constables July 5, 2005)

In the above examples, women's definitions point to both coerced sex as well as gestures and experiences within the workplace that lead to difficulty in executing one's duties. More importantly, their words reflect both the daily-ness and everyday-ness of these advances. By "daily-ness" I am referring to the repeated nature of this unwanted interaction, and by "everyday-ness" I mean to suggest the "normalcy" with which these gestures, acts, and requests are perpetrated. This dissonance in defining sexual harassment does not simply mark a difference of perspective between men and women in the RBPF. Rather, it marks a fundamental disjuncture between those who have the power to enact rules and regulations regarding sexual harassment and the capacity of those most affected by the absence of these rules to convince those with power of the legitimacy of their experience.

It is important to note, though, that the need to convince those in power is not a conversation among equals or between individuals. The discourse around sexual harassment throughout the region is such that "culture," ephemeral and chameleon, is unleashed to protect national territory from "encroaching" ideas. In post-colonial societies ever wary of neo-imperialism, this is a profoundly convincing argument. As such, the culture of Barbados is both implicated and deployed in further de-legitimizing claims towards naming sexual harassment as harm, as evident in the following responses:

> Well, I have had problems with that particularly in the American context. For example, if you call everything sexual harassment, well, how then do you approach a female if you like her? Is that sexual harassment? If I say to you you're a good looking woman, is that harassment? And listening to Americans in Barbados, that has not been considered as sexual harassment. You have to set a definition that sets what should be interpreted as sexual harassment.
>
> Any act for which a female feels uncomfortable is sexual harassment, and this is our problem now in cultures because when I look at America and England, a person making a pass, sometimes compliments can be misconstrued as sexual harassment. From African descent, now, we don't understand that. The problem is that we are so Europeanized, we don't even know that we're black sometimes. (interviews with male police constables July 19, 2005)

There are a number of important features evident in discussions of sexual harassment and culture in the Caribbean. First, sexual harassment is deliberately misnamed and, by extension, trivialized: it stops being repeated, unwanted sexual advances and becomes an obstacle to adults who wish to express *and receive* intimate intent. Second, discussions pertaining to sexual harassment are rendered as a North American import and, relatedly, addressing sexual harassment serves to undermine some sense of Caribbean "authenticity" i.e., the way in which men and women do things in "this culture." This argument implies, however, that women do not know how to read their own culture or interpret inappropriate expressions of interest. Regardless of the rationale, culture here is deployed as a platform on which male power is able to perpetuate its authority. It does this by using culture to censor what women *can* legitimately name as harmful (Tamale 1995, 702). In other words, the use of culture craftily silences women by naming sexual harassment as antithetical to Caribbean society. Women who challenge "culture" are then seen as betraying the "norms" of their culture. This culture, though, has been named in ways that secure men's sexual privilege and access to women or other marginalized sexualities.

EXPERIENCES OF AND RESPONSES TO SEXUAL HARASSMENT

Gladys Brown-Campbell's brief study of patriarchy in the Jamaica Constabulary Force offers a fifty-year overview of women's progress and constraints within the Force. A detective sergeant at the point of writing, Brown-Campbell observed that:

women who repulse male advances which amount to sexual harassment and are "punished" for what is termed their "impudence," they [sic] are given consecutive night duties, their names are left off promotion lists and they can be transferred to remote areas because they "dare" to think "that they are God's gift to man" as one spurned former officer stated. (1998, 27)

It is not accidental that this insight comes from an "insider." In a very early discussion about sexual harassment, MacKinnon argued that women were likely to underreport experiences of harassment because of shame (97). Within para-military organizations, there is also the fear of retribution and punishment when the ranks close off and move in on the victim.

Despite the absence of formal reporting channels within the RBPF, three things indicated that sexual harassment existed in the RBPF: (1) women's narratives of experiences of sexual harassment, (2) women's stated techniques of circumventing sexual harassment, and (3) institutional masculinized narratives about sexual harassment. The hierarchical nature of the RBPF is such that "quid pro quo" forms of sexual harassment by male senior officers toward junior female officers cause the most concern. The following re-crafts the narrative given by a female Special Police Constable within the RBPF:

> X was unemployed and encouraged to become a special police constable by Inspector Y. She successfully completed her training and during the course of executing her duties became intimately involved with a male police constable. She was later approached by Inspector Y and told that it was because of him that she was able to get into the RBPF, and if anyone should have got sex from her, it should have been him. This was followed by Inspector Y grabbing at her on at least two occasions. (recrafted narrative from interview with police constable July 12, 2005)[14]

This was the most glaring example of *quid pro quo* sexual harassment rendered throughout the research. Male power in this narrative does not only derive its currency from the officer's rank, but there is an attendant aspect of patronage. Patronage preys on women's ignorance of the bureaucratic system, as in this case, where the candidate's access to the force supposedly hinged on the influence of the Inspector. The corollary is also true, and it is that that the narrative of patronage preys, too, on women's knowledge of the bureaucratic system and their awareness of the system's unresponsiveness to their needs, hence, the need for male patronage.

The "daily-ness" and "every-dayness" of sexual harassment was most evident in the fact that all the women interviewed had devised a coping strategy or coping mechanism to buffer themselves against sexual harassment:

> Where I am working now, they are there and people make remarks about it, but that is men. That is that person's ambition. But for me as a woman, I would not put it up. I am not offended. I came there and met it.
>
> Well, you have to know how not to get too friendly because, for me, I have my line, and once you cross it, that's it for me. It's your mouth, you could say what you want to say. My personality is that I feel sorry for a person like you who think that to speak to you in such is a way is going to get you to respond. But if they ever lay a hand on me, I will do something about. And if you use your position to try to oppress me, I might be working here, and I not paying you any mind, and you look to transfer me or picking on me, that is something that I would have to deal with.
>
> Someone might use their office to get something. It might not actually happen, but they will try to intimidate you. And you give in to that pressure. (interviews with female police constables July 5, 2005)

These excerpts suggest the prevalence of sexual harassment and the women's recognition that sexual harassment is a somewhat intrinsic aspect of being in an atypical profession ("I am not offended. I came there and met it"). The female officer's sense of having met sexual harassment within the RBPF is both true and untrue. In the immediate sense, she is right. She did come in to the RBPF and found an environment that facilitated sexual harassment. However, in a more long-term sense, sexual harassment emerges *in the context of* an increasing presence of vulnerable sexualities in the workplace. Prior to women's presence in the force, men in the main did not sexually harass each other. It did not exist prior to the increasing numbers of women in the force. Rather, it is an exercise of power *in response to* the increasing numbers of women in the force.

Women, therefore, exhibit resignation; that is, they play along in the hope of becoming one of the boys. Alongside anticipating sexual harassment as inevitably part of the work environment, women use both verbal and non-verbal gestures to demarcate personal space. Women identified a number of strategies of dissuasion: silence (stopped speaking to the individual), humor, gestures, and posturing to ensure that sexual advances were clearly recognized as unwanted. Yet, these tactics merely positioned women defensively in an antagonistic environment without exacting any behavioral adjustments from male members of the force.

On the other hand, the institutional narratives render a very different story of sexual harassment. In this story, men are seduced, and women are temptresses:

> This is not just, this is applicable to the RBPF but not confined, but there are situations where females are attractive. They do not have stripes, but they might use the power that the person has . . . their attractiveness.
>
> I believe that a lot of women, well, there is mainly men in the force, and I believe that there are a lot of affairs and that that softens things up.
>
> I must admit that we have not dealt with gender, and some of the supervisors are not as forceful in dealing with their female subordinates. I don't mean to be obnoxious, but women can get favors and that some men are seduced into behaving in that manner.
>
> I think the force is suffering from chivalry. We men think that women are soft and put them in cushy jobs and so, and they, too, reach out, too, by their overtures. (interviews with male police constables and members of the Senior Command Team July 5, 2005)

Admittedly, sex can function as currency for women if and when debarred from organizational sites of power. That this becomes one of the few available options for women to navigate bureaucracy is itself a form of sexual harassment. However, what is striking in these narratives is that when called upon to firmly address sexual harassment within the RBPF, formerly authoritative men lose all sexual agency and re-write themselves as duped, besotted, and victims of women's seductive attractiveness and overtures. Thus, sexual harassment becomes a contemporary problem because the organization now has women.

Labeling women as lesbians has also emerged as a strategy within the RBPF to deflect attention from men's culpability as harassers. Labeling non-compliant women, regardless of their sexual identity, as lesbians is not a new tactic aimed at diminishing woman's relevance. Suzanne Pharr, in her article "Homophobia as a Weapon of Sexism," identifies lesbian baiting as

> an attempt to control women by labeling us as lesbians because our behavior is not acceptable, that is, when we are being independent, going our own way, living whole lives, fighting for our rights, demanding equal pay, saying no to violence, being self-assertive, bonding with and loving the company of women, assuming the right to our bodies, insisting upon our own authority, making changes that include us. (1988, 260)

To label a woman as lesbian with derogatory intent prefigures the marginalization and stigmatization that is to come. In contrast, lesbians as gender non-conformists occupy a liminal space within non-traditional professions and can be seen simultaneously as peer, threat, and competitor. Elizabeth Hoffmann, in her study of female cab drivers, for example, explored the ways in which women used lesbianism, regardless of their sexual identity, to avoid sexual harassment (2004, 7).

In addition to men's vulnerability to women's seductive power, sexual harassment is positioned squarely at the feet of women as perpetrator and victim:

> I don't know if it's blown out of proportion, but there is the sense that they are recruiting a high percentage of lesbians or somehow they may come with those tendencies and get converted. When it's this topic of sexual harassment, the men will say, it ain't the men that interfering with the women, is the women interfering with the women. It's just stuff that I hear since I don't be around so many females. I think that one is frowned on more. (interview with member of the Senior Command Team July 17, 2005)

Or, one officer, before exploring his understanding of "sexual harassment," prefaced it with the caveat:

> Before I start, though, sexual harassment that I am aware of has been transgender and within the gender.
>
> *Interviewer:* So, you're saying same sex and male/female?
>
> Well, we have to look at all the possibilities. I would not stick my neck out and say that it is not present, male/female. I would suspect that that is most prevalent. (interview with male detective July 17, 2005)

The masculinist culture of the RBPF, infused as it is by heterosexist discourses, succeeds on two counts in its deployment of lesbianism. It deflects attention away both from men's roles in the occurrences of sexual harassment and the masculine culture of harassment. Further, with only 2.5 percent of RBPF's strength accounting for women above the rank of constable, it is unlikely that these acts of *quid pro quo* harassment would manifest, presenting us with an interesting conundrum where, if such acts were to occur, women would be held to a different conceptualization of harassment than men hold themselves. At the core of this discussion, though, is the desire for gender conformity, where as Franke noted earlier, "sexual harassment (becomes) a technology of sexism." To this end, lesbian baiting serves to keep women in line and men off the hook.

WOMEN'S INSTITUTIONAL EXPECTATIONS OF THE ROYAL BARBADOS POLICE FORCE

The previous section began with the circular predicament that within the RBPF there are no formal grievance procedures to address sexual harassment specifically and, as a result, there are no reports of sexual harassment. However, as many officers noted, even if there were formal rules and reporting regulations in place, usage would remain low because of fears of reprisals:

> We have the procedures in the force. But if it happens, you can adopt the procedures for redress, but they also know the fear of victimization and that fear of victimization is greater than anything else in this force. Sometimes, it might not even be true. But it was perception if a person go and report, and Inspector then others of similar rank will close ranks and shut you out because it used to be so. It used to be that way. The older women, they will tell you back in the sixties and the seventies, the officers used to have a field day with them. (interview with Sergeant July 18, 2005)

Nonetheless, female officers oscillated between healthy skepticism to outright disbelief that their reports would be taken seriously if formal reporting procedures were in place:

> *Interviewer:* Would you report it?
> *Respondent:* Of course, I would do it. Nothing may be done, and it might backfire on me ... And that is another thing, too, you won't find officer going against officer. You have to find somebody all the way up to the top on your side.
> *Interviewer:* Do you think they would take it seriously?
> *Respondent:* I can't say I know because I don't see very much happening. You might not be given certain opportunities.
> *Interviewer:* Sexual harassment is dealt with fairly by the force if reported?
> *Respondent:* Well, if it goes in writing, everybody will try to get rid of it and pass it on. So, from that standpoint, yes. So, you will find at the end of the day, justice is served.
> *Interviewer:* Do you think a report of sexual harassment will be taken seriously by the administration?
> *Respondent:* Well, it's just like a rape case. The victim is going to be expected to come forward and state the matter. (interviews with female police constables July 5, 2005)

The final respondent's allusion to a victim's interrogation within a rape case is astute. The absence of any rules and regulations to demand behavioral change of perpetrators as well as the invocation of culture to minimize the significance of harassment leave women needing to defend themselves in the midst of the perpetration of harm. The women's responses indicate that the RBPF faces an increasing loss of institutional credibility regarding sexual harassment. The women expect to be circumvented, penalized, and in need of a senior benefactor for their reporting of sexual harassment to be taken seriously.

How are these messages conveyed within the RBPF? The RBPF's complacency regarding sexual harassment significantly contributes to this sense of institutional distrust. The following phrase reflects the basis of this distrust:

> Yes, there are reports, but people don't like to go through the procedure. I think they do it just to let you know, let the individual know or let the organization know that I have nothing to do with him or nothing to do with her. We do get reports, formal and informal. (interview with member of Senior Command Team July 15, 2005)

In the above citation, there is a disjuncture between the report and the action taken in response to this report. There is a disavowal of any punitive action for reports are made merely "to let the individual know . . ." As a result of these silences, perpetrators are essentially aided and abetted by the RBPF in acts of sexual harassment against marginal sexualities in the workplace.

As gender minorities increasingly find themselves in previously closed work environments, the need for preventative and protective mechanisms will become even more urgent. Through my examination of masculinist organizational cultures, I attempted to show both the overt and subtle practices of inequity. Access is never enough. Without the appropriate protections, access may offer little more than trauma with income. Caribbean legislators have begun the work toward the provisions of such protections, but slowly, rarely, and, as this chapter's case study of the RBPF demonstrates, the remedies are inadequate. However, my overall concern has been the disjuncture that exists between mainstreaming debates and sexual harassment legislation. The imperative of this chapter, therefore, has been to broaden mainstreaming debates in ways that expand the debates—from technical approaches to mainstreaming to a consideration of what we do next on behalf of those bodies that have been "mainstreamed."

6 Development and Identity Politics
Securing Sexual Citizenship

> ... by using abjected populations as exemplary of all obstacles to national life; by wielding images and narratives of a threatened "good life" that a putative "we" have known; by promising relief from the struggles of the present through a felicitous image of a national future; and by that, because the stability of the core image is the foundation of the narratives that characterize an intimate and secure national society, the nation must at all costs protect this image of a way of life, even against the happiness of its own citizens.
>
> —Lauren Berlant,
> *The Queen of America Goes to Washington City:*
> *Essays on Sex and Citizenship*

In this chapter, I pour libations to the memory of Michael Sandy. Michael Sandy was killed in October 2006[1] after connecting with his assailants in a gay Internet chat room. They lured him to a parking lot in Plum Beach, New York, beat him, and chased him onto the beltway where he was struck by a car.[2] After five days in a coma, Sandy succumbed to his injuries. I read of Sandy's death in the *New York Times* over a cup of coffee one crisp fall Sunday morning. Against the serenity of the morning, this report began as one more hateful tragedy until I read that his aunt was MaCartha Lewis. Because amnesia is the thing that allows me to live away from "home," the name prompted me to call my cousin to confirm why this name was so strangely familiar. "Isn't 'MaCartha Lewis' Calypso Rose's real name?" I asked her, and she, being the archivist of all things ancestral, replied, "Yes, who died?" I then proceeded to quickly tell her that someone from "home" had been killed in New York. Sandy was the nephew of a very prominent female calypsonian. So, it follows that I was disheartened to realize that Sandy's attack and passing was not reported in any of the local dailies in Trinidad and Tobago. I remember him here.

In many ways, this chapter is precisely about why it is important to re/member Sandy's murder. Where did he belong, and which nation should we have expected to mourn his passing? To think of citizenship through the lens of mourning seems appropriate for a discussion of sexual citizenship but not enough. When called forth to dialogue with each other, "sexual≠citizenship" signals multiple forms of dispossession, and as Judith Butler notes, these dispossessions give rise to other meaningful relations in which grief becomes a "resource for politics" (2004, 19–23). In this chapter, I want to think about the ways in which such political resources are used by individuals who occupy non-normative sexual identities that allow them to see themselves and to become meaningful to and recognized by the frameworks of governance, development, and social life. So, although I begin my discussion of sexual citizenship as an experience of loss, I hope to end by means of struggle where the use of sexual citizenship comes to signify an explosion of what Berlant discusses as the "cruel and mundane strategies (that) promote shame for non-normative populations and ... deny them state, federal and juridical supports because they are deemed morally incompetent to their own citizenship" (1997, 19). Building on my earlier notion of embodied equity, I center a queer subjectivity,[3] not for the purposes of inverting a hetero/homo binary, since such a tactic leaves power unscathed. Rather, in keeping with Gayatri Spivak's characterization of deconstruction, I center the marginal in order to notice (1999, 373). This act of noticing demands that the margins become active so that we can work within and through the margins. Such a centering "in order to notice" helps us better understand the terms that create the idea and materiality of marginality. Such a centering also helps us to complicate and nuance the singular flatness on which the very terms of marginality depend (ibid., 175). To this end, in this chapter, I continue to be guided by my use of the term embodied equity as a way of seeing through the body. I want to explore an emerging body of literature that is forcefully challenging the excision of the queer body from discourses of development.

From here, I re-position my vantage point to examine the ways in which sexuality rights talk has been gaining ground globally through the dissemination of the Yogyakarta Principles (Yogyakarta) in 2007, designed as they are to bring human rights considerations to sexuality.[4] I aim to satisfy my curiosities about the Principles by placing them in conversation with existing literatures on sexuality in the Anglophone Caribbean. By way of imaginative flight, I willfully speculate about what the existing literature might tell us about the im/possibilities of the Yogyakarta Principles in relation to the pursuit of equity in the region. Yet, since this is an imaginative flight, I return in order to critique the dominant representation of the Anglophone Caribbean as always-already homophobic in ways that make the life suggested by the Yogyakarta Principles unimaginable. In no way dismissing the egregious hostilities of the region, I nonetheless aim to make an argument for what I refer to as sexual survivance by exploring the ways in which embodied equity is being pursued by the Society Against Sexual Orientation Discrimination (SASOD) in Guyana.

EMBODIED EQUITY AND THE TERRAIN OF RESPONSIVE DEVELOPMENT

To use the word "sexuality" in relation to the Caribbean enacts a tropological association that generates a series of contentious terms and subject markers such as "exotic," "wanton," "homophobic," "hypersexual," and "macho." It is not my goal here to disentangle this mess of meaning; the associated meanings are there and can only be accounted for. Neither do I focus on the ways in which sexuality is embedded within other pressing sex-related phenomena in the region, such as domestic violence, sex work, trafficking, or heterosexual gender performances. Rather, I want to think very specifically about sexuality in relation to questions of development with a view to examining how this interface allows for a Caribbean that is more responsive to calls for embodied equity.

Feminist insistence that "gender" is something we do rather than a possession or something we have has radically challenged contemporary gender binaries and their related determinisms (Lorber 1994; Fenstermaker and West 2002). However, as if a cautionary tale to the risks of interchanging analytical categories as though they were commensurate, applying a similar set of assumptions to discussions of "sexuality" leads to drastically conservative tendencies of sexual fixity within queer discourses. That is to say, if sexuality is defined only as a set of behaviors and activities, this can potentially become woefully prescriptive. It becomes prescriptive in the sense that we then presume to know the list of appropriate behaviors that should be aligned with a given identity, but more importantly, those who do not perform the designated behaviors of their assigned identity become suspect.

Admittedly, behaviors and practices are important to the social relations of sex. Additionally, the sanctions applied to sexual behaviors and practices are instructive as powerful indicators of contemporary biopolitics. On the other hand, defining sexuality via behaviors and practices too readily comes to rest on what we do and allow to be done with our genitals—a critique, while still sadly salacious, Eve Sedgwick reminds us, is inadequate (2008, 29). To this end, Judith Halberstam notes that "in our frenzy to de-essentialize gender and sexuality, we failed to de-essentialize sex. The analysis of sexual practices does more than simply fill in the dirty details; it also destabilizes other hierarchical structures of difference sustained by the homosexual/heterosexual binary" (1998, 114). Here, Halberstam acknowledges the subversive force of doing sex non-normatively. However, equally as compelling is that even when we point to particular sexual practices that are similar, we should resist the inclination to interpret them as the same (118).

With these caveats in mind, I am using "sexuality" to refer to the historically and culturally contingent nodal point of ideology, behaviors, identities, and practices as they come to rest on different bodies. Because the locations I have named are co-constituted, I find it difficult to explore questions of sexuality as an emphasis on behaviors without appropriate attention to the fact that what these behaviors mean is determined by historical, ideological, and individual political deployments (Weeks 1990).

While practices of governmentality would have us believe that there are easy alignments among embodiment, behavior, and identity, my use of sexuality dislocates these assumptions. Sexuality in this project functions as an axis of power with the capacity to repress or enable expressions of self and structures within the social order. My use of terms such as "lesbian," "transgender," and "gay" points to examples of these expressions. I am particularly interested in a consolidation of "sexuality" that helps us to understand decision making in relation to sex, desire, and pleasure, and I am using the term to facilitate a critical engagement with the belief systems and regimes that push individuals out of or into hiding around sex, pleasure, and un/belonging within Caribbean societies. With this approach, it is not enough to simply say, as is often said within the African diaspora, that sexuality is not a category by which individuals make sense of themselves in the world. The historical contingency of sexuality, rather, points to sexuality as always-already at work within subject formation in modern society, and silences are more often indicative of trenchant heteronormativity as opposed to sexual irrelevancy (Foucault 1990).

Though a nascent area of study, the connections between sexuality and development have become more evident and, when inflected through the prism of human rights, urgent. The absence of sexual minorities from considerations of development brings the field of development and its many agents (state-managers, development practitioners, academics, NGO workers, and advocates) into an insidious complicity with the very antithesis of its "intent"—discrimination. This discrimination may be marked by the exclusion of same gender-loving people from community/village projects (Swarr and Nagar 2004). It also includes the epistemic erasure of sexual minorities or the pathological representations of queer subjects that makes other forms of discrimination possible and desirable (Gosine 2005).

Such disciplinary exclusion has also prompted some to note that the field has proceeded apace without a body (Cornwall et al. 2008). These critical engagements observe that when the body does appear, it is shaped by an assumed heteronormativity in which gender, sex, and sexuality are in neat and un-conflicted alignment.[5] They point further to the ways in which this heteronormativity establishes legitimate actors and recipients of development and guides the allocation of resources and the shape of policy (Swarr and Nagar 2004; Cornwall et al. 2008). I would argue, moreover, that the erasure of sexuality and the body from the field of development is another instance of the influence that neo-liberalism's paradigms have had on the contemporary shape of development studies. As such, development projects continue to share neo-liberalism's faith that women's market participation will alleviate their poverty and that matters of embodiment are irrelevant unless they pertain to violence or similar factors that may impede market participation. Within this framework, sexual minorities become neutered workers, hence, the field blinds us to the very specific forms of discrimination that LGBT individuals experience once incorporated into the market.

By invoking a sense of embodied equity, I position the body as a dynamic socio-historical formation that challenges these attempts to circumnavigate the material body. Embodied equity acknowledges that it is the physical body that mediates the social world and, further, that it is on differently situated physical bodies that regulatory narratives write their words. The body, therefore, becomes a way of bringing into relief three simultaneous moments in my discussion of equity—the first moment points to a recognition of these regulatory powers, the second demands accountability from these regulatory schemas, and the third potentially allows us to envision a world where these things (if painful) are no longer done.[6]

By way of process, embodied equity in the context of development demands a greater comfort with sex talk, in all its permutations and manifestations. In their piece "Power and Desire: The Embodiment of Female Sexuality," Holland et al. survey 150 young women in the United Kingdom about their interaction with their sexual identity and sexual performances. The authors examine the socially mediated difficulties that constrain young women's sexual expression (1994, 22). They found that the young women in their study used a range of rhetorical strategies to perform the role of the "good sexual subject." These strategies, all practices of veiling, resulted in what they referred to as the young women's "sense of detachment from their sensuality and alienation from their material bodies" (ibid., 24). I want to extrapolate from the connection that Holland et al. make between talk and disembodiment to think about what their arguments would mean if applied to a wider social collective. The social taboos and silences that exist around certain kinds of sex similarly work to excise and tame the body within public discourses—as become evident in my parenthetical interventions in Holland et. al. in the following discussion:

> Talking about what women (Lesbians?) do with their bodies (Pleasure each other? Inflict pain?) exposes and threatens the carefully social constructed disembodied sexuality (Within law? education? parliament?) (parenthetical questions mine; Holland et al. 24).

My very deliberate parenthetical interventions show the ways that sex talk, when inserted within a wide range of arenas, destabilizes and challenges the politics of disembodiment. However, a few caveats are worth repeating. First, my use of sex talk should not be seen as positing the truth of sex; rather, I am threading multiple practices of sex through power with the hope that unexpected and surprising forms of sex talk enact Halberstam's call for the de-essentializing of sex. Also, my use of embodied equity does not aim to ground or fix an essential location. By pointing to the materiality of the body—and queer bodies specifically—I am aware that I have placed a socially constructed embodiment into a holding pattern. This holding pattern, however, is productive because it allows us to notice (what has been done), to re/member that regulatory mechanisms mark flesh and to envision alternative practices of engagement.

To notice, to re/member, and to envision bodies and queer bodies specifically has been glacial within existing development discourses. Nonetheless, the literature on development as an embodied practice has already begun to initiate a very timely interrogation of the unmarked assumptions within the field. Such an interrogation is not only conceptual but has also produced a dynamic and exciting interdisciplinary conversation in which the seams of development studies are being undone by queer theorists, much in the same vein that feminists and women's studies scholars challenged the field of development to drastically re-imagine its boundaries some four decades ago.

This emerging literature insists that the body and sexuality are not new terrain for questions of development. It does so in two ways: first, by laying bare the field's own assumptions about proper bodies and their place; and, second, by showing that bodies, sex, and sexuality are already implicated subsets of the areas seen to be the legitimate areas of concern for development (e.g. housing, water, and jobs). Susan Jolly, for example, has offered a resounding critique of the development practitioner's preoccupation with a body that can only be imagined as a recipient or perpetrator of "bad sex" (2007). Without trivializing what she identifies as the actual dangers of "bad sex," that is to say, the danger of rape, domestic violence, or HIV/AIDS, Jolly points out that accompanying development's preoccupation with danger has been the elimination of any consideration of pleasure, and further, that this "bad sex" analytic requires the deployment of age-old gender stereotypes for its effectiveness—that is to say, women remain victims of sex, and men continue as the perpetrators of bad sex (2007, 12).

Jolly's attention to pleasure, if incorporated analytically, has the capacity to produce a cataclysmic re-orientation of the field, cataclysmic in the sense that it complicates our understanding of how the body navigates the murky interface of danger and pleasure. Can women, for example, choose sex work without having to contort their decision to fit with poverty as the economic (read appropriate) justification for that choice? It is also cataclysmic in the sense that, by following Jolly's logic, it is not enough to merely stop the things that allow bad sex to occur, but rather, the attention to pleasure dictates an additional step that re-frames the policy environment so that individuals are able to move toward the achievement of pleasure. Finally, development practice turns on many problematic presumptions about what people of the global South deserve. Attention to pleasure "fleshes" the humanity of the citizens of the global South where questions of pleasure are seen as worth policy aspirations.

While Jolly's work shows that a conceptual narrative about sexed bodies has always accompanied development discourses, other scholars in this emerging literature offer a complementary discussion of the ways in which bodily considerations, whether acknowledged or not, are already at work within the traditional terrain of development. Henry Armas, for example, contests the tendency to position considerations of sexuality as secondary to considerations of poverty (2007). Armas's discussion is one that shows the economic costs of inhabiting atypical gender locations as a way of connecting sexuality to the infrastructure of development. For Armas, the economic vulnerability that results for pregnant teens or feminine boys who are both coerced to leave school as a result of being socially ostracized is equally as important to the achievement of their life chances as more mainstream poverty alleviation strategies (e.g., skills training programs).

Armas's arguments, while strategic, remain dangerously instrumental. Strategically, Armas points to sexuality's indivisibility from questions of material well-being and makes it clear to practitioners that the very project goals they desire stand in jeopardy if sexuality is disregarded. On the other hand, Armas repeatedly returns to the "economic consequences" of ignoring "sexuality," thereby leaving his analysis open to a degree of economic instrumentalization. Although Armas argues that sexuality should not be seen as secondary to questions of poverty, the category's legitimacy within Armas's framework is predicated on arguing for sexuality's economic relevance; in other words, it makes good sense/cents to address sexuality. In such a model, addressing sexuality as a category garners salience in relation to a series of productive arguments (e.g., risky jobs, creation of poverty, and income insecurity) that not only invokes a neo-liberal framework, but in so doing, is ironically unable to achieve the author's desire to displace sexuality's secondary status to things economic.

However, Armas's work also represents a third intervention by this emerging body of literature. Building on earlier integrations of a human rights-based approach to development, this third intervention applies a sexuality rights-based critique to governance approaches within development. Development-based sexuality rights advocates, framed by an earlier declaration of "women's rights are human rights," and rights-based reforms to development within the UN, have been able to draw on a mode of reasoning that is increasingly familiar to development studies.[7] By arguing for the importance of legal limits to and redress for violations, as well as the moral suasion that resides within terms such as participation, accountability, dignity, and rights that a rights-based approach provides, sexuality advocates have been able to apply the primary strengths of a rights-based approach.

The United Nations' "Human Rights-Based Approach to Development Cooperation: Towards a Common Understanding Among the UN Agencies" (the Stamford Statement of Common Understanding) is now foundational to more contemporary iterations of a rights-based approach within development, significantly influencing the language that sexual rights advocates deploy. The Stamford Statement of Common Understanding offers sexuality advocates three significant points with which to ground their arguments within the field of development. First, the Stamford Statement allows sexuality advocates to draw on the almost sacrosanct underlying philosophy of the Universal Declaration of Human Rights. Second, the Stamford Statement offers sexuality advocates important implementation aspirations that call for the integration of human rights considerations into all agreements, technical instruments, and policies. Finally, it enables sexuality advocates to argue for the importance of capacity building so that "duty-bearers" are able to meet their obligations and of "rights-holders" to claim their rights.[8] Taken together, this rights-based turn with development allows sexuality rights advocates to build on prior discourses within the field in ways that encourage traditional development practitioners and technocrats to see sexuality's "integrality and indivisibility" with prior development goals and also demands some consideration of the ways participation for the integration of sexuality rights would refashion the development project itself.

The Stamford Statement's call for complete integration of a human rights approach to all stages of development policy and planning brings to mind the Nairobi *Forward-Looking Strategies* articulated in 1985, which called for a comprehensive consideration of women as a category of analysis within the UN. With such an antecedent, were sexual rights activists to deploy similar language in relation to sexual minorities, it would consequently read that governments should aim "to involve and integrate ~~women~~ sexual minorities in all phases of the planning, delivery and evaluation of multisectoral programmes that eliminate discrimination against ~~women~~ sexual minorities, provide required supportive services and emphasize income generation." This re-framing captures the re-visioning that characterizes the third invention of the emerging literature on sexuality and development.

Finally, this emerging body of literature offers a place-based empiricism of sexuality that brings a more comprehensive understanding of the ways in which my previous conceptual conversation matters to real bodies in real spaces within the field of development. This intervention itself is a treasure trove of inter- and multi-disciplinary scholarship, with insights from sociologists, queer theorists, public health specialists, development practitioners, literary critics, and anthropologists alike. Amanda Swarr and Richa Nagar in their piece "Dismantling Assumptions: Interrogating 'Lesbian' Struggles for Identity and Survival in India and South Africa" question the general presumption in development theory that sexuality is a phenomenon whose importance belongs to the North, or the privileged. This, they argue, not only reflects development theory's own constructed binaries but also serves to "facilitate the belief that no poor lesbians live in the South" (2004, 494). Admittedly, "the poor" have been among the primary ways in which development has been able to see the South. However, if we see sexuality rights as important to development without needing to postulate economic justifications for such considerations, is there then a place in development for sexual minorities who are not poor in the South?

Provocation aside, Swarr and Nagar's (2004) attention to the relevance of sexuality as an analytical category within the global South helps us envision the expansive possibilities of applying a sexuality rights-based approach to development. Swarr and Nagar's work, for example, shows how the presumed heteronormativity of development sanctions the exclusion and disciplining of gender variant bodies by development practitioners, NGO workers, and grant recipients, a matter that could be preemptively addressed if the non-discrimination of sexual minorities were integral to the terms of accessing donor funds, monitoring, and evaluation. Place-based empiricism also offers an important caution to the universalizing tendencies of rights-based approaches. In this vein, the fact that two of the women in Swarr and Nagar's study live together as married but do not identify as lesbian offers an important and timely critique against the imposition of Northern understandings and terminology regarding sexual expression as well as teleology of sexual becoming.

Each of the four interventions that I've outlined as characteristic of the emergent literature allows us to see the limitations that the field of development has imposed on the body or the spaces that have been pried open for the body's expansive possibilities. With each intervention, we have a different insight into how the lived body has been positioned or should be re-positioned within the field of development. Whether the body is seen as narrowly sexed (e.g., reproductive or vulnerable), as excluded, as enmeshed (e.g., within housing and transportation policy), or as an inalienable rights-holder, what is clear is that the interventions presented by this mushrooming literature continue to irreversibly transform the field.

This literature's transformative potential notwithstanding, it runs the risk of remaining mere potential unless accompanied by the support of institutional frameworks that are meaningful to practitioners. This demand for institutional endorsement leaves development practitioners and sexuality rights activists caught in a competing pull with potentially very different end points. Whereas the disciplinary interventions I have pointed to previously emphasize the importance of place-based specificity and nuance, the international approach to rights and development seems to be unduly susceptible to universalizing homogeneity, a lack of institutional locations through which actions can cohere, and a fractured, conservative political landscape on issues of sexuality. In other words, even as the emerging literature aims to transform the field, the field will not go willingly. It is not necessarily important to reconcile these two seemingly disparate spaces; however, learning how to navigate them is critical for effective political advocacy.

Despite the best of intentions, the UN institutional framework is an unpredictable location for gay rights advocacy. LGBT activists have much to celebrate in the UN Human Rights Committee (HRC) ruling in Toonen v. Australia in 1994, in which the HRC expanded the understanding of discrimination on the basis of sex to include sexual orientation. On the other hand, it is now well documented that the International Lesbian and Gay Association (ILGA) has had more than a contentious relationship with the UN ECOSOC in their repeated bid for recognition of consultative status (Waites 2009; Mertus 2007).[9] Similarly, while the "Statement on Human Rights, Sexual Orientation and Gender Identity" was read by Argentina with the support of sixty-six states of the 192 states of the General Assembly, it is hard to forget that the relatively innocuous Brazilian Resolution supported by nineteen countries five years earlier faced resounding and successful opposition from five nation states and the Vatican.[10]

These halting gains, in addition to a powerful conservative lobby within the UN, undermine any teleology of "sexual progressiveness." Whereas for the category of "gender," documents like the CEDAW name and consolidate a set of legal best practices to which countries might aspire toward in order to mitigate all forms of gender discrimination and provide a set of guiding principles with which states can re-imagine their statehood, no such document exists within the UN framework for sexual minorities. It is for these reasons that the Yogyakarta Principles has so forcefully filled a gap within the international arena of rights for sexual minorities.

Crafted by a twenty-nine-member team of internationally diverse high-level human rights experts, the Yogyakarta Principles succeed where the UN has not in its articulation of a "universal guide to human rights which affirm binding international legal standards with which all States must comply." The twenty-nine Principles center the State as having a duty to uphold and protect against discrimination based on sexual orientation and gender identity. Rapporteur for the development of Yogyakarta, Michael O'Flaherty, notes that twenty-nine principles are first ordered to establish the fundamental human rights of all persons regardless of sexual orientation and the right to be recognized before the law in protection of these rights. This foundation in turn supports issues of bodily integrity, namely the right to life, security, privacy, and justice. They call for the elimination of discrimination based on sexual orientation and gender identity within the socio-cultural and political domains of life (e.g., employment, housing, and education). Additionally, states are expected to eliminate constraints and limitations on thought, expression, and movement and to call for the protection of those individuals who work on behalf of or defend the rights of those who are discriminated against based on sexual orientation or gender identity. Finally, Yogyakarta calls for redress for those who have been harmed and accountability from those who have harmed on the basis of sexual orientation and gender identity (*Yogyakarta Principles* 2007; O'Flaherty 2008).

The framers of the Yogyakarta Principles, evidently mindful of the UN's institutional significance in international human rights debates, have called for its endorsement by the United Nations High Commissioner, the UNHRC, and ECOSOC, as well as complete integration into the programming, treaties, and offices of the UN. The Yogyakarta Principles additionally consolidated its desired connection with the operations of the UN by timing its launch with the general session of the UNHRC on March 2007 (O'Flaherty 2008, 237). The Principles also re-articulate and affirm earlier human rights sentiments and principles resonant in documents such as the Universal Declaration of Human Rights, CEDAW, and the Convention on the Rights of the Child. However, by centering the vantage point of sexual minorities, Yogyakarta expands the possibilities of subject-hood for a range of different and unexpected individuals.

Yogyakarta is to be commended for recognizing and protecting the sexual identity of young children. The HIV/AIDS crisis gave children a new visibility

Development and Identity Politics 181

as the absorptive surface of the HIV/AIDS crisis. For example, the maternal/ child transmission risk pointed to the need for humanitarian medical aid, the myth of the female virgin curative showed the gendered nature of childhood risk, and children also emerged as caregivers to younger siblings due to parental loss as a result of HIV/AIDS. Recognizing the limitations that children experience regarding self-advocacy, they feature as a protected category in the Yogyakarta Principles with regard to medical abuses and alterations (Principle 18), homelessness and eviction (Principle 15), access to resources regardless of their or their parents' sexual orientation/gender identity (Principle 13), that sexual orientation/gender identity of the child or the child's parents ought not to be positioned as incompatible with the best interest of the child (Principle 24), and, most importantly, that children in accordance with their maturity level should have the right to express their sexual orientation and gender identity (Preamble). Taken together, the Principles re-imagine childhood as an important site of sexual identity formation that takes account of risks but also acknowledges a set of rights that warrants protections from the state.

Yogyakarta takes on a protective and assimilative mantle. It pursues a wide range of objectives spanning from protections against the most egregious violations based on sexual orientation and gender identity (e.g., torture and unlawful detention) to the everyday matters that render life meaningful and dignified for many contemporary subjects (e.g., housing, education, and modes of association).

Skillful drafting notwithstanding, Yogyakarta is supra-institutional, not state endorsed, and non-binding. Consequently, they face vulnerabilities that are similar to those discussed in relation to gender mainstreaming approaches and others that are unique to Yogyakarta because of the specific forms of discrimination manifested against sexual minorities (Waites 2009). The principles, as with the terms of gender mainstreaming, articulate visions and best practices that establish meaningful parameters for dialogue between sexuality rights activists and state agents. Gender mainstreaming's advantage, of course, has been the integration of its rhetoric into other documents adopted by many of the world's countries (e.g., Beijing and CEDAW), a gesture which surely raises the stakes (though not enough) on questions of accountability in a way that is yet to occur for considerations of sexuality. However, the effectiveness of the Yogyakarta Principles, as for gender mainstreaming, hinges on the creative ways in sexuality rights activists are able to strategically deploy their language and philosophical underpinnings in their everyday political work.

SPECULATIVE VERNACULARIZATION: THE YOGYAKARTA PRINCIPLES AND THE ANGLOPHONE CARIBBEAN

In the absence of formal legal state-based endorsements, Yogyakarta depends on sexuality rights activists and state-managers who already share its goals for its circulation and integration into a global legislative framework. The nature of sexuality rights activism in relation to the Principles is such that the terms of the Principles have to be rendered meaningful and intelligible within a set of varied and disparate local contexts, many hostile to its objectives. As I have discussed in terms of gender mainstreaming, the labels of "alien" and "inauthentic" have functioned very effectively to delegitimize the call for rights for marginalized subjects. Additionally, the exclusionary repertoire that is the rhetorical strategy "threat to national security" has been used increasingly in a post-9/11 militarized world to further limit which citizens get to be part of the public discourse on rights. It is into such pernicious contexts where activists are constantly called to weave new arguments in their demand for equity and justice that anthropologist Sally Merry inserts her idea of vernacularization.

Merry's use of vernacularization marks moments and practices of adoption and adaptation within local contexts and is integral to the effective border crossing of "human rights language." In this process, a diverse body of intermediaries works to re-locate, to interpret, and to integrate the transformative intent of a human rights framework. Merry's idea of "vernacularization" is an active process that traces and initiates how ideas are "adapted to local institutions and meanings" (2006, 39).[11] Her idea of vernacularization speaks to the set of actions and strategies that activists *do* in relation to adapting a human rights discourse. However, I want to put her idea of vernacularization to slightly different purposes that are more speculative in intent.

I want to think about the adaptability of the Yogyakarta Principles to the Anglophone Caribbean using existing literature on sexuality as the intermediary of vernacularization. This is an important conversation that delineates the two frames of reference that sexuality rights activists will need to negotiate in their human rights work in the region—namely, what we already know and the world that we envision. Further, this mode of speculative vernacularization helps us to examine some of the limitations of Yogyakarta when brought into dialogue with the Caribbean context. What light does the literature on sexuality in the Caribbean shed on the im/possibilities of the Yogyakarta Principles and conversely how might Yogyakarta push us to envision a different reality to what exists presently in the region? That the assimilative aspect of the Yogyakarta Principles is needed as a means of transforming the sexual rights landscape of the Caribbean is evident from the concerns raised within existing literature.

Yogyakarta brings with it a human rights framework with all of the philosophical underpinnings of a rights-based agenda. So, while it is appropriate to laud a rights-based approach for its attendant principles of indivisibility and inalienability, it also produces a homogenizing effect, and the transmission of an individualist ethos.

My characterization of Yogyakarta as assimilative warrants some explanation. The Yogyakarta Principles gets its legitimacy and recognition from its reaffirmation of already endorsed iterations of human rights-based assertions. This familiarity contributes to its assimilative tone; however, it is also assimilative in the sense that it accords sexual minorities the right to belong to the everyday of their respective worlds (e.g., housing, education, culture, and public association). Yogyakarta is transformative in the sense that the "we" of the everyday have never had to imagine sexual minorities as having a right to these spaces and such envisioning disrupts the terrain of the everyday.

The heteronormativity into which the Yogyakarta Principles would potentially intervene in the Caribbean has been well established from a number of different disciplinary vantage points. Jacqui Alexander offers one of the earliest critiques of Caribbean states' disciplinary force on the body. Her work promoted serious consideration of the stratagems by which Caribbean states work to control women's sexual agency (1997), to constrain women's access to the portals of same-sex pleasure (1991, 1994),[12] and to manipulate women's sexual autonomy in its own quest for futurity (1997). In the interest of embodied equity and development, what, then, might be the interface between this landscape and a rights-based approach, such as the Yogyakarta Principles?

There is an aura of "timelessness" that attends rights-based discussions, which Yogyakarta imports through articulations such as inalienability, dignity, and freedom (Preamble). Indeed, much of the moral authority invested in human rights-based approaches to development depends on this very aura of timelessness: timelessness suggests that there is a historical worthiness residing within personhood that warrants the application of rights. I am not suggesting "timelessness" in the sense that contemporary rights can be applied to earlier societies. As Yogyakarta shows by naming sexuality in the contemporary period, a rights-based approach must constantly adapt and expand the criteria by which human/rights are understood. However, the aura of timelessness remains persuasive as a kind of temporal totality. I'm drawing on the dialectical sense of temporary totality, which Sartre describes as that period that "totalizes before becoming partial" (2004, 60). Applied to my discussion of the aura of timelessness, I am suggesting that the established philosophical underpinnings of rights convey a *sense* of timeless for the duration of their urgency—that as a construct it represents a sensibility and a moral worth that belies their actual years of articulation.

Alexander's work halts this sense of timelessness. Her work may be seen as a reminder to sexuality rights activists that the verncularization of the Yogyakarta Principles in the Caribbean must take seriously that sexuality and sexual subjects emerge within and over time and confront the implications of historical antecedents. Alexander's use of time in her work arguably holds important lessons for sexuality rights activists which I would like to explore here. First, it centers the effects of colonial time on "modern" subject positionings. Here, she observes that the subjugation of sexual minorities is a crucial component of modern state-managers' attempts to forge a counter-discourse to the colonial rendering of the savage sexual other (1994).[13]

Second, from Alexander's discussion of colonial time we may extrapolate that any sexuality rights intervention, such as the Yogyakarta Principles, should be cognizant of the ways in which colonial trauma translates itself into sexual discrimination. My investments do not rest with linear narratives of historical trauma, but rather, the understanding that the vernacularization of sexuality rights is always already in dialogue with residuals of a colonial gaze that has produced different sexualized and racialized subjects. Tracy Robinson has critiqued the tendency to presume that colonial law has proceeded relatively unchanged on questions of family and conjugality (2009). My point here is a different one. I am noting that the colonial project was in part an erotic project that deployed racial/ethnic and gendered categories toward the production of hypersexualized, effeminate, or wanton subjects (Alexander 1994; Stoler 1995). The residual effects of this colonial erotic project shadow contemporary performances of sexual locations. As a result, gay subjects are inflected by racialized antecedents with distinctly different gender performances at their disposal.

Furthermore, the re-appearance of colonial trauma garbed as nationalist resistance to the presumed neo-imperialism of sexuality rights matters greatly to sexuality rights terrain. A number of powerful structures find their resting place in the silencing incantation of "cultural authenticity" as that which is antithetical to non-normative sexual locations. It is almost hallucinogenic to observe state-managers, who, with one wave of the cultural wand, attempt to make sexuality and sexual minorities disappear. This rhetorical gesture oscillates between the desire for an imagined heteronormative past and nationalistic distancing from the "North."

Take, for instance, a debate that occurred in the Trinidad and Tobago Parliament on the potential repeal of the death penalty, which rapidly reaches for the trope of sexual treason. In the exchange that follows, the then oppositional People's National Movement representative for Diego Martin East, Hon. Colm Imbert extended a congratulatory note to Amnesty International for its advocacy against the government's reintroduction and enforcement of the death penalty. The Minister of Trade and Foreign Affairs, Hon. Mervyn Assam, in response, warned the nation against Amnesty International's "nefarious" intent, arguing:

> They want to Europeanize everything. They want to turn us into homosexuals. The European Parliament wants to pass legislation forcing countries that depend on them for aid, grants, or technical expertise, and so forth, they want you to pass legislation so that homosexuality could become law. [Crosstalk] Yes, could become law; but they reduced it in England, under 16 years, because England is now part of the European Community. They are forcing dependent territories; those British

territories that are still colonies, they are forcing them to introduce legislation into their colonial legislatures to bring in homosexuality as law. You have to be very careful. ("Finance Committee Report," *Trinidad and Tobago Hansard* September 22, 2000, 908)

The slippage from a conversation about the death penalty to state sanctioned homophobia (They want to turn us into homosexuals) is as an example of the compulsion to protect the heteronormativity of the state (Alexander, 1994). However, I take Cynthia Enloe's analytical caution in which she highlights the idea of the nation as an embodied phenomenon: the "nation never speaks" unless enabled to do so by a wide assortment of state functionaries. If the nation requires bodies to speak, then the ease with which state-managers can harness the idea of homosexuality in the fissures of a conversation about an unrelated matter such as the death penalty suggests that a more micro-social phenomenon is it work.

These slippages announce the heteronormativity of the state; however, they also position the idea of homosexuality as a fetish for state-managers. In this sense, (male) "homosexuality" erupts as an irresolvable personal anxiety for heterosexual (and some homosexual) state-managers, and as we see in this previous parliamentary exchange, is ritualistically (rhetorically) revisited in order to displace one's crisis of identity onto the fetish. As Anne McClintock suggests:

> The fetish marks a crisis in social meaning as the embodiment of an impossible irresolution. The contradiction is displaced onto and embodied in the fetish object, which is thus destined to recur with compulsive repetition . . . By displacing power onto the fetish, then manipulating the fetish, the individual gains symbolic control over what might otherwise be terrifying ambiguities (1995, 184).

McClintock's observations applied to the parliamentary exchange help us understand how the homosexual fetish is caught up in a range of state fears regarding neo-imperial impositions. In this sense, the perceived non-procreating homosexual not only threatens the state's desire for futurity but enacts yet another betrayal by opening state-managers to international scrutiny and judgment.

The above exchange offers an example of what Alexander refers to as the palimpsestic effect and the ways colonial time might be felt in relation to sexuality rights and development. Thinking of sexual subjects as bodies in post-colonial time gives pressing urgency to the ways in which nationalist rhetoric invokes colonial hauntings and anxieties about imperial impositions as justification for the marginalization of sexual minorities.

Alexander's engagement with time in her work also disrupts the march-of-sexual-progress narrative.

This disruption of time is simultaneously a disruption of place in so far as the march of progress accords modernity to the North. Here, Alexander notes:

> Since there is, analytically at least, no good heterosexual democratic tradition and no bad heterosexual primitive tradition, there can be no false deduction that democratic heterosexualization is simply more benign in its alignment with modernity than traditional heterosexualization which, in its alignment with backwardness, is simply more pernicious. (2005, 194)

Despite the aura of human rights' timelessness, what I've attempted to show here is that the vernacularization of sexuality rights in development is potentially riddled with history as an ambivalent present. As such, we need always ask what do rights mean in the context of history, that is to say, what are the historically derived narratives that have been produced to secure ongoing sexual subjugation as well as what are the fictions of subjecthood, that have been produced out of these narratives?

Both time and geography are indicted in Alexander's above critique. The complex sexual cartography of the Anglophone Caribbean, the prevalence of homophobia as here *and* there defies the simple binary that Alexander critiques. Part of the Caribbean's complexity is that its sexual cartography, due to larger migratory patterns as well as internal configurations, is far more fluid than internal and transnational practices of border control acknowledge. There is certainly a need for greater research on the intraterritorial control (and defiance of such) that aims to determine when and where queer subjects might enter within the respective countries of the region. The practices of space/place making of queer geographies within the region, the strategies by which space and place are claimed, are of great interest to me, particularly in light of the region's density and the proximity within which social life is lived.

However, in keeping with our ongoing interrogation of the Yogyakarta Principles in relation to sexuality rights and development, I am simultaneously heartened and concerned by the fact that Yogyakarta reinforces an individual's right to reside in a location where she/he is able to "do and be" regardless of sexual orientation or gender identity. By calling for the right to asylum, Yogyakarta reinforces the importance of realizing one's capabilities and right of mobility and migration to places where realizing such capabilities is possible (Principle 23). This is an appropriate and proper position, which, in practice, becomes messy.

There are many immediate instances that challenge and complicate the Yogyakarta Principles' hint of a suggestion of safety as elsewhere. Let us take, for example, two starkly different yet analytically similar moments. The first returns us to Michael Sandy's assault in New York, which I outlined at the beginning of this chapter. The second moment is a brief interview exchange in which Anderson et al. examined the narratives of ten men who have sex with men (MSM), all from the Caribbean diaspora (more precisely Jamaica, which brings its own methodological fallacy of conflating nation with region), (2009). The interviews explored the respondents' expression of internalized homophobia, practices of self-surveillance, passing, and where possible, a comparative discussion of how their experience of their sexual identity differed when in the Caribbean and the United Kingdom. The exchange that I find intriguing is found under a subheading entitled "Policing Homosexuality." The interviewer asked a Caribbean-born respondent:

> *Q:* Were you on the DL [down-low]? To which he responded:
> *A:* It's not like you could be anything else in Jamaica! (C35)

The interviewer went on to observe:

> According to this man, all gay men in Jamaica are forced to perform a public heterosexuality to satisfy the demands of heterosexist society. Here, he was not referring to the commonly understood meaning of down-low, where MSM engage in concurrent heterosexual and homosexual relationships: he meant that the relationship had to be discreet.

Michael Sandy's attack in the United States, juxtaposed against this anonymous interviewee's categorical denial of any sexual possibilities in the Caribbean, undoes facile binaries, particularly those that would obscure the ways in which the intersections of race, nationality, and sexuality also position some sexual subjects as abject within the North. That these two situations emerged and were experienced across locales provides an interesting opportunity to complicate the multiply situated lives of sexual subjects from and within the Caribbean (I hesitate to say queer) diaspora.

Transnational feminist critiques have long taught us to read for the imbrication of identity and place, as well as the ways in which issues, activism, and policies "becomes global and how the global is evident in the local" (Desai 2005, 320). From such a perspective, borders cannot represent any one thing. They may lock people in or provide the means of escape. They may be ephemeral through technological or cultural transfers, fixed through modes of policing, or come to one's door in barrels, visiting relatives, and remittances. They may be adobe walls or the zinc of a tenement yard, country/rural divides or class/ethnic divides. A border may be a blink so that someone who loves you accords you the privacy of not seeing. Borders are, in this sense, liminal. Queer subjects navigate all of these border terrains to some degree.

Amar Wahab and Dwaine Plaza provide an excellent discussion of these multiple vectors at work in the context of the Canadian-Caribbean queer diaspora and the ways that borders are made and remade as migrants struggle to belong to new spaces and inhabit multiple subject positions (2009), as does Rinaldo Walcott's discussion of the ways that a black queer diaspora complicates the "local, national and transnational desires, hopes, and disappointments" of nationalist projects (2005, 92).

Liminality, however, does not place borders outside of power, but it does fracture presumed knowledges that travel these sexual circuits—such as in our interview exchange above, the ease with which our interviewer presumed an unmediated flow of meaning in the concept "DL." The interviewee's understanding of "DL" was distinctly different from that of the author's and fractured the author's presumption that there would be a "common understanding" of the term and that it would be used by the respondent to structure his life in a particular way. Thinking transnationally rather than in the binary of here/there helps us to map the ways in which the concept "DL" travels differently and encourages us to pause long enough to identify the kinds of questions that are produced from this difference. For example, what might this differentiated understanding of DL tell us about how the "closet" functions differently in small island societies? By mapping subtle flows of sexual formations, embodied sexualities also makes suspect any desire to render geography as homophobic. The flows that are central to my offering of embodied spatialities also make it difficult to articulate seamless enunciations, as our anonymous respondent does, that nothing else was possible but to be on the DL.

Similar gestures of transnational fracturing are at work in Jasbir Puar's (2001) discussion of drag diva performances in Trinidad. She maps circuits of desire and the performance of drag as a disruption of U.S. heteronormativity, hegemonic whiteness, and imperialism. However, when she eventually interviews two of the drag performers and they short-circuit her idea of gender identity, we see, in the fractures of the text, a glimpse of a life where sexuality is integral to identity but deployed to purposes that are distinctly at odds with the interviewer's as they are disappointed that Puar has no leads to help them break into the U.S. market with their drag performance.

What do these considerations of borders, circuits, and fractured diasporic sexuality mean for the vernacularization of the Yogyakarta Principles? Understandings of safety as elsewhere may in some cases be appropriate and in others, not. Mobility does not always secure the safety of queer subjects, and safe travel is not always possible for queer subjects. Travel here encapsulates an understanding of how borders (both internationally and nationally) are policed along a range of other identity locations in ways that limit the ability of some queer subjects to travel in the sense of changing location. These constraints may include deportation (permanent movement from North to South), the inability to afford or qualify for a visa (movement from South to North), or the much understudied lack of desire to leave whatever locale may be understood as "home."

The vernacularization of sexuality rights in the course of Caribbean development also requires some engagement with rights-based approaches as spaces of inordinate conflict (Darrow and Amparos 2005). In the Caribbean, this conflict manifests in very different and sometimes unexpected ways. Because sexual rights signal a redistribution of power, we expect there to be the attendant struggle over the criteria by which rights may be claimed (ibid.). However, extant literature also suggests interesting modifications to the language of sexuality rights that impinge on the effective vernacularization of the Yogyakarta Principles in the region.

Rights-based agendas, such as the Yogyakarta Principles, lend a framework of legal and moral legitimacy to the issue under discussion. For many advocacy groups, invoking the idea of rights offers a shortcut to a common understanding of infractions and ideas of appropriate redress. However, as David Murray's research in Barbados shows, protections that are premised on a group's special status such as sexual orientation may not always be seen positively or as progressive. Murray recounts the public debates that followed in Barbados after then Attorney General Mia Mottley in 2003 called for the repeal of sodomy laws from Barbados's Criminal Code. In the public fora that followed, Pat, a member of the United Gays and Lesbians Against AIDS in Barbados, declared that after having been repeatedly ostracized and discriminated against in everyday public spaces, he had begun to work with the UN Declaration of Human Rights to assert his right to equal treatment. Murray subsequently asked Pat why he chose the Declaration of Human Rights, noting its silence on sexual orientation, to which his respondent replied that what was needed in Barbados was not gay rights but human rights (269). In other words, Murray's respondent rejected the special rights status of gay rights as a platform for his politics.

At first gloss, Pat's rejection of special status may read as a fairly conservative take on sexual advocacy. However, in the context of my speculative vernacularization of the Yogyakarta Principles, I want to explore why an LGBT activist might reject a sexuality rights agenda and think a bit about the possible repercussions for the Yogyakarta Principles.

Broader theoretical concerns about the efficacy of identity politics in part explain the possible rejection of rights based advocacy premised on the special status of sexual orientation. Among these critiques has been the presumption that an identity-based platform leads us to presume to know the political stakes *a priori* of context. As such, both identity and the associated political agenda can become fixed, essential, already known. This presumption that we can know both identity and political agenda outside of context then offers a facile alignment of politics, identity, and behaviors. To reject a sexuality rights-based approach premised on such alignments disrupts all attempts to sanitize identity, behavior, and politics and, interestingly, opens up possibilities for even a heterosexual location to be more than it presently is, what Moya Lloyd refers to as a subject-in-process (2005). I am not suggesting that important political work cannot be done in the name of an expressed identity.[14] The women's and civil rights movement make this clear. However, these movements have had, to different degrees, to deal with various challenges to the cohesion of their framing identity markers.

Yogyakarta proves to be quite promising in this regard. Despite its emphasis on questions of sexual orientation and gender, the terms "lesbian," "gay," "bi-sexual," "transgender," "transsexual," and "intersex" appear once in the document. The Principles are activated rather by the categories "sexual orientation" and "gender identity." Ostensibly, on these grounds, one need not be locked into a singular articulation of identity. The Principles, guided as they are by these terms, maintain some degree of validity in territories where the organizing terms LGBT are not necessarily the organizing categories by which sexual expression occurs. For instance, while not focused on the Anglophone Caribbean, Gloria Wekker's groundbreaking study of matism in Suriname offers such a refusal of sexual identity and in so doing holds the tyranny of sexual fixity at bay; a matter that continues to confound my students for whom LGBT resonates as the site of politics (2006).[15]

Matthew Waites, however, cautions against a hasty embrace of the Yogyakarta Principles' use of the terms "sexual orientation" and "gender identity." The heteronormative frame of existing human rights discourses contributes to Waites's contention that the categories of sexual orientation and gender identity have been integrated within an international framework without adequate problematization. Drawing on Judith Butler's idea of a "heterosexual matrix," Waites asserts that the categories "sexual orientation" and "gender identity" garner their cultural intelligibility through a referential relationship with heterosexuality. Building on Sara Ahmed's work, he argues that "orientation" requires one to position (or orient) one's self to something else, which leaves the concept as operationally limited as the categories "LGBT." Waites further argues that the increasing medicalization and psychologizing (processes that themselves are unduly informed by what people do with their genitals) of the categories "sexual orientation" and "gender identity" render them vulnerable to another rendition of fixity due to the discourses of truth making within scientific domains. Finally, he contests the Yogyakarta Principles' assertion that sexual orientation and gender identity are universal and integral to everyone's "dignity and humanity" (2009).

While these are vital considerations, from a transnational perspective, I maintain that the categories of sexual orientation and gender identity as deployed by the Yogyakarta Principles offer a wider range of political possibilities to sexuality rights activists in the field of development. Though the categories LGBT imply a directed and/or named direction to which one should orient, Yogyakarta attaches neither directional nor numerical coordinates to its definition of sexual orientation. One may choose to orient or not (hence, room for those who do not center sexuality as critical to subject formation, dignity, or humanity), and one may choose to orient, or not, to one other person or to many others (Yogyakarta Preamble). It is precisely this lack of specificity that gives the Yogyakarta Principles a subtle radicalism, that is to say, it allows for a transformative LGBT politics by a different name.

However, I would hazard a guess that none of my foregoing discussion appropriately accounts for why Murray's respondent rejected a gay identity politics as the grounds for his rights-based claims. I would posit rather that Murray's respondent possessed more than a perfunctory understanding of the extent to which the categories LGBT have been overdetermined by stigma and shame in the Caribbean in ways that can potentially limit the political efficacy of the categories LGBT.

"Morally deserving" rights holders have to be made; these are not natural locations. These categories are made and remade in ways that reinforce the state's notion of the deserving rights holder. To this end, Andil Gosine's exploration of the ways in which public health discourses have remade gay men as culturally and politically intelligible primarily through the lens of HIV/AIDS may be helpful in our discussion of why an activist might reject a gay rights platform (2005). Gosine rightfully calls for a degree of skepticism about the ways in which sexual minorities are framed within the war against HIV/AIDS in the Anglophone Caribbean, particularly because sexual minorities have acquired some degree of "visibility" through the HIV/AIDS pandemic. This coupling, therefore, as Gosine notes, positions the gay body as something to be managed in the interest of public safety (ibid.)

His work traces the remaking of gay men as diseased subjects and captures the seductions of funding imperatives which bring gay activists into complicity with these remakings of identities. In his later work, we see that part of this remaking hinges on the deployment of the category MSM. The term medicalizes and, under the shadow of HIV/AIDS, only allows for a conceptualization of same-sex as something that is dangerous, dangerous not only to themselves but also to the heterosexual collective since these are men, the narrative goes, who are closeted. Consequently, this overdetermination forecloses on naming other identity frames as legitimate frames by which to claim rights—can the body that struggles with HIV/AIDS still claim the right to pleasure? Such a question becomes pivotal to Gosine's critique of sexual intelligibility through HIV/AIDS. One of the worrying conflations that occurs through the coupling of gay activism and HIV/AIDS is an association of male same-sex activity with ideas of shame, stigma, and, as Gosine observes, pathology (2005, 2009).

The overdetermination of LGBT locations through stigma and shame in the Caribbean raises pressing concerns about how one argues the rights of sexual minorities. It is here that Yogyakarta begins to feel tautological. In other words, how do we claim human rights for those whom the state refuses to see as human? And here we have the most difficult impasse of the Yogyakarta Principles, as Thoreson notes, "If rights cannot actually be safeguarded for unpopular groups, the legitimacy and moral foundation of human rights are called into question" (2009, 324). With such overdetermination at work, one understands the rejection of sexuality rights through the label of gay rights, but nonetheless, such overdetermination should also point to why sexuality rights activism in the region requires an aggressive capture of the symbolic field within which statements are to be enunciated about sexuality, development, and equity in the region.

SYMBOLIC CAPTURE: DISCIPLINING SEXUALITY

This aggressive capture of the symbolic field—or what I point to as symbolic capture—refers to acts and practices that challenge the perceived coherence of dominant narratives about sexuality in the Caribbean. Gestures and practices of symbolic capture render intelligible issues, identities, and spaces that practices of governmentality would have us not see or see as perverse. In this sense, symbolic capture functions counter-discursively. It also functions as a mode of resignification of what might be said, among other things, about bodily practices, spaces of un/belonging, and the myth-making tools of law and politics.[16] The urgent need for this symbolic capture is connected to the vitriolic nature of the already-said as the "hollow that undermines from within all that is said" about sexuality and Caribbean subjects (Foucault 1972, 25).

Jenny Sharpe and Samantha Pinto (2006) argue that the study of sexuality in the Caribbean has not only been taboo but also essentially off limits for wider scholarly research. However, categorically declaring the study of sex and sexuality taboo and off-limits runs the risk of simplifying the many ways in which we do come to understand something about sexuality from the seepages and fissures of a range of diverse texts—i.e., as part of the incorporeality of the already-said. Symbolic capture is a practice of reading against the grain, or reading symptomatically, the diverse arenas that generate their own spaces of sanctioned inattention with regard to sexuality, whether sexuality is a *primary category of analysis or not.*

The growth of masculinity studies in the Caribbean has questioned the constructed nature of masculinity and has interrogated various forms of male privilege inclusive of sexual privilege. However, a symptomatic read, for example, of Barry Chevannes's ethnographic study *Learning to Be a Man* (2001) shows the intrinsic nature of sexuality to a range of disciplinary and methodological discussions. Chevannes's three-territory (Dominica, Guyana, and Jamaica) study highlights the heterosexualization of masculinity through his examination of the rites and rituals of performing heterosexuality that are critical to how men learn to be men. Sharpe and Pinto, in their literature review, commend Chevannes for the ways his work addresses the complex sexual relations that exist between homo/heterosexual formations and for the fact that his study on masculinity presents a "welcome break from research that takes heterosexuality as the invisible norm, even if homosexuality finds limited space in his formulation of the "street" and "yard" culture" (2006, 266). However, read symptomatically, the "limited space" is itself instructive and presents an example of how the sanctioned inattention of mainstream disciplines simultaneously disciplines homosexual identities. Homosexual identities emerge as significantly less than a "limited space" in this text. They are all but erased or emerge in the context of negation.

Chevannes, in a discussion of the "tolerance" that existed in one of his ethnographic sites for a particular homosexual man in the community, writes:

> According to informants, homosexuals like "Lady Jane," a male sixteen-year old high school student, are tolerated in the community. Because they were born and grew up there, people have learned to accept them and do them no harm, provided "dem never inna fuss wid any body." The only harassment they get is a "little jeering" or a "tax levy" on those who serve as whores (2001, 203).

Centering a queer subject in order to notice, I am struck by the fact that we never *hear* Lady Jane and how s/he names herself or himself (as opposed to the derogatory appellation of the respondents); we do not understand the form that the "little jeering" takes and how s/he deploys certain performances, probably around masculinity, to navigate these tense moments.

On the other hand, we find the caricature of the man-hating lesbian evidenced by "Man fi dead! Mek uman run things! My life change from mi stop worship man believe me!" (ibid., 202) Where we see queer subjects, they signify as non-speaking, or we find an acritical representation of lesbianism as failed female heterosexuality. Might there be room for questions to be asked about desire, self-identification, or gender performance? My concern here is that as development practitioners and scholars engage the "field," it is imperative that we question the texts that we draw on for our "grounded" understanding of sexuality since "new" research can rewrite "old" phobias.

Practices of symbolic capture are always in contestation with practices of foreclosure, even those that implicate one's own activism agenda. So, while I've gestured toward disciplinary imbrications where queer representation hovers as a practice of negation, I am keen on the increasing representation of the Caribbean, by both conservative and progressive forces, as the most homophobic place on Earth. I am not concerned about whether this immeasurable articulation is so or not. I think about the politics of such superlative language and examine the extent to which such language may actually serve as a strategy of containment for sexuality rights activism in the Caribbean.

The confluence of political opposition to gay travel to the region (Alexander 1997; Puar 2002), the widespread and rapid circulation of cultural products expressing homophobic sentiment (Chin 1997; Saunders 2003), the international activism mounted against these cultural products (Chin ibid.; Gosine 2009), the continued presence of religious fundamentalism

(Gutzmore 2004), a small but growing number of asylum petitions based on sexual orientation,[17] and the prevalence of HIV/AIDS in the region which comes to rest on gay bodies in problematic ways (Gosine 2005), have informed the symbolic representation of the region as superlatively homophobic.[18] That the region has not always been experienced or imagined this way is evident in the interstices of existing scholarship. Murray notes that the queer respondents in his study voiced an increasing sense of hostility directed toward them when compared to two decades ago, for example.

In "Man-Royals and Sodomites: Some Thoughts on the Invisibility of Afro-Caribbean Lesbians" (1992), Makeda Silvera vividly brings to mind her early childhood memories, with a narrative voice almost whispers in our ear as she says:

> I heard "sodomite" whispered a lot during my primary school years, and tales of women secretly having sex, joining at the genitals, and being taken to the hospital to be "cut" apart we were told in the schoolyard. Invariably, one of the women would die. Every five to ten years, the story would surface. At times it would even be published in the newspapers. Such stories always generated much talking and speculation. (1992, 522)

Silvera goes on to say that only later in life was she able to understand her grandmother's biblical admonitions to her when she first came out and that her grandmother's warnings resulted from a deep awareness that "any woman who took a woman lover was indeed walking on fire . . ." Silvera continues, "I began to see how *commonplace* the act of loving women really was, particularly in working class communities" (emphasis mine; ibid., 532–534).

I highlight Silvera's observations because they provide some insight into the ways in which myth making, regardless of its religious or infantile origins, works to erase the presence and complexity of queer subjects—and also because her memories insist on what she refers to as the "commonplace act of loving." Wesley E. Crichlow, on the other hand, tells a more angst-ridden tale of his own coming of age as a "bullerman" [sic] in Trinidad and Tobago. It is important here to acknowledge that differentiated social and legal sanctions that exist against male same-sex intimacy cannot be understood outside of a broader discussion of maintaining male heteronormative hegemony to which homosexuality signals affront, instability, and betrayal. However, one feels the narrative sigh and playful mischief of his appropriation of "bullerman" as Crichlow writes about the deployment of certain calypsos for same-sex entertainment and the discovery of "bullerman" spaces. From Murray's interviews and Silvera's and Crichlow's stories, we know that despite conservative efforts within the region to identify LGBT persons as foreign, without history, and dangerous, the presence of lesbians, gays, and transgender people is neither a new or discontinuous phenomenon.

Development and Identity Politics 195

In this light, what are the politics of homophobic superlatives? While I identified the circulation of homophobic cultural products as only one component of a confluence that led to the region as superlatively homophobic, I want to pause briefly to suggest that in the early 1990s something significant happened (Williams 2000; Walcott 2009). In 1992, dance hall now conscious vibes-megastar Buju Banton, released "Boom Bye Bye," a trenchant, caustic war cry that called for the death of gay men. This song generated a number of responses that have been well outlined by cultural and literary critic Timothy Chin, who documents that the song was banned on a number of radio stations, concerts were canceled after calls to boycott were issued by LGBT advocacy organizations in Europe and North America (e.g., Gay and Lesbian Alliance Against Defamation and Gay Men of African Descent), calls came for an apology, and the song was discussed and disavowed in a number of U.S. mainstream publications (e.g, *Vibe, Billboard, Village Voice, New York Post*; Chin 15).

Banton's defenders immediately responded to these attacks with much of their defense being mounted on the holy grail of culture. Scholars such as Carolyn Cooper, Chin notes, chided the opposition to Banton's song as an uncreative lack of imagination. Banton's lyrics, Cooper argued, were not to be interpreted literally. This was a defense, Chin notes, that was echoed by African-American hip-hop feminist Joan Morgan, who argued that U.S. aversion to Banton's lyrics was a "misunderstanding of Jamaica street culture."

While I am identifying Banton's "Boom Bye Bye" as a significant event, I am not offering it as an originary narrative; I am not suggesting that Banton's "Boom Bye Bye" causes homophobia, which is often what is argued in the literature. Rather, as Michel Foucault suggests, I want to make use of Banton's song "just long enough to ask myself what unities are formed" as a result of "Boom Bye Bye's" eruptions. I want to accept Banton's "Boom Bye Bye" as "significant" long enough to pause to see what processes were legitimated in the politics of the superlative.

The first is that Banton's "Boom Bye Bye" signaled a timely *consolidation of rather than creation of* homophobic sentiment.[19] The irony of the debate was that widespread circuits of disavowal in mainstream magazines also provided Banton with a circuit of consolidation, i.e., the very forums within which there was a refusal of Banton's ideas provided, by their very retelling, a rhetorical repetition and, as a result, inscribed the idea of the region as homophobic.

Further, Banton's lyrics were never designed to rest within the national as a geopolitically confined construct. Rather, it very successfully facilitated a diasporic call to action in his "shout-out" to the "New York crew," "Brooklyn crew," and "Canadian crew." This has received very little analytical attention in the literature. However, British filmmaker and cultural critic Isaac Julien provides us with a bit of insight into the dispersed sense of black gay diasporic anxiety that resulted from Banton's "Boom Bye Bye." Talking about his own experience in London, he observes that this fear "was the experience of a lot of people, Black gay and lesbian friends who lived in Black areas like Brixton in London. I would feel a general tension if I were to go to the London Annual Black Carnival, now more than ever, for example" (1995, 409). I am using diasporic anxiety here to think about the ways in which migrant and migrating bodies and well-being are at times traumatically caught up or implicated in location/space and the multiple circuits through which culture travels. So, it becomes necessary for Julien to negotiate his sexuality in a way that has to take into consideration that something has occurred elsewhere in the diaspora but nonetheless marks his body. It also allows us to challenge the belief that flows of identity-making are unidirectional North–South. The larger point is that the migrant/migrating body is always in dialogue with multiple modes of reception and self-making.

Banton's "Boom Bye Bye" provided beat and rhythm for existing homophobic sentiment. As with much of hip-hop, dance hall, and reggae, there is a rhythmic seduction that requires the listener to do deliberate political work to challenge the lyrical content, to resist the seduction. "Boom Bye Bye" consolidated a genre of music that could be encapsulated under the theme of bun fire/more fire (Saunders 2003). While redemption, retribution, and fighting the woes of Babylon have always been a part of the subversive element of reggae, within dance hall this bun fire/more fire thematic introduced a religious brimstone theology with a clear delineation of sinners and judges that immediately added to the popularity of artistes such as Sizzla, Capleton, and Elephant Man.

Most importantly, however, the attention that "Boom Bye Bye" received enacted the twin processes of hypervisibility and overdetermination. The processes of hypervisibility (where one thing is excessively seen) and overdetermination (where one thing stands in for a range of other things or comes to mean several different things), regardless of where enacted, always produce myopic vision. These twin processes of hypervisibility and overdetermination have been most instrumental in the designation of superlative homophobia. Jamaica became the entire region; so, there was an overdetermined conflation at multiple levels that made it impossible to see difference within Jamaica and with Jamaica. The idea of the region as homophobic was further consolidated by repetition and hyperbole. The credibility of these superlative narratives increased with support from "authentic narrators" who participated in these narratives as part of their own political and, at times, entrepreneurial practice.[20]

My recurring caveat is not to suggest that the Caribbean is not hostile toward LGBT subjects. It is also not even to be relativist and suggest that there is homophobia everywhere since I agree that degrees of scale matter. I am more concerned about what we are unable to see and do when we, as LGBT activists, theorists, and scholars participate in these overdetermined narratives of homophobia. Rather than furthering the cause of gender equity, our participating in these narratives has us participating in the same practices of erasure that virulent homophobes do. To declare the region homophobic, even when used within progressive organizing, depends on the same kind of logic as does the politically conservative declaration that there are no homosexuals here.[21] As I've argued earlier, both declarations are ahistorical; both erase the presence of LGBT subjects; and both merely describe or declare without an engagement with civil society.

What does it really mean when an entire spatial geography, its cultural practices, and its peoples are labeled as homophobic? I want to cautiously suggest that it is a label that conceals significantly more than it reveals. The term constructs a generic Other that conflates homophobic geography as identity, a statement that describes both people and place without any allowance for exceptions. It tells us something about the environment within which sexual minorities live, but it does not tell us *how* sexual minorities live. The designation, as often imputed from and by the North, invariably invokes the tensions of imperial and colonial histories. It blinds our ability to see instances of gay life and gay activism at work within these societies and to see formations of community. If we look only at homophobic spaces and materials, for example, repressive legislation, policy, or popular media, then gay men and lesbians only emerge in forms of negation, that is to say, what is not possible.

Most importantly, this superlative representation leaves only one location from which sexuality rights activists are able to speak—that of victim, which feminists have learned is a complicated but limited positionality. A region's representation as superlatively homophobic has become so overdetermined within popular, academic, and political discussions of the region that, I want to suggest, the lives of sexual minorities are further disabled as they only emerge under the sign of the region's homophobia. Conversely, sex talk, symbolic capture, and agency as re-signification open spaces for engagements that are not framed only by victimhood. These approaches grasp the strategies by which the state makes this the only viable location from which to speak and simultaneously work to make marginality a political choice to be deployed if deemed strategically effective. It bears repeating that I am not dismissing the variegated hostilities of the region. My goal is also not to simply state that queer activism exists in the region, though that is in itself an important and transformative statement for those who have not been able to envision such a possibility. I want to position and to notice ongoing practices of symbolic capture as a means of suggesting that rights should be claimed from positions of strength and entitlement and to situate this activism as indicative of a challenge to modes of overdetermined foreclosures that exist presently.

A PRELIMINARY FORAY INTO SEXUAL SURVIVANCE: SOCIETY AGAINST SEXUAL ORIENTATION DISCRIMINATION

My call for practices of symbolic capture is designed to construct a grid of sexual survivance that allows us to learn from the strategies of sexual rights activists who challenge forms of erasure and terror and provide a better vantage point from which to see the policy, legislative, and social responsibilities of the state toward minority communities.

My use of the term survivance draws from Native American scholarship and the work of Gerald Vizenor, who offers us survivance as a way of discussing Native Americans' push to go beyond mere surivival. His use of the term captures ideas of self-definition, of social, sacred, and political awareness, as a well as a sense of history while looking forward to the future (1994). For Vizenor, survivance is "an active sense of presence" (2008, 19). In the context of sexual rights advocacy, I further define sexual survivance as research, literature, and scholarship that privileges place, practice, and personhood of sexual minorities toward the realization of life, community, and well-being. Sexual survivance, therefore, challenges the rhetoric of tolerance as it is not possible without a political commitment to an engaged struggle of symbolic capture. In other words, actively engaging the literature, research agendas, policy and legislative landscapes, and social worlds that would rather not see is the only way that sexual survivance occurs.

Despite and probably because of the problematic and axiomatic rendering of the Caribbean as "homophobic," we have to begin to interrogate how sexual minorities make life (as opposed to survive) in the midst of hostile conditions. This is a preliminary gesture toward a queer geography that presents a cartography of belonging and place-making through a re-signification of what the state wishes to make of queer subjects in Guyana.

In this light, there is a growing and vibrant sexuality rights advocacy landscape in the Caribbean.[22] While Crichlow postulates that a change in the Caribbean's sexual landscape will inevitably result "due to North American hegemony," I maintain that this viewpoint robs us of the richness of the sexuality rights landscape that in a very preliminary and cursory way I would like to make visible.

Society against Sexual Orientation Discrimination is a Guyanese NGO whose primary mission is law reform and the expansion of the visibility of sexual and gender minorities in the Guyanese cultural, political, and legal landscape. SASOD began in 2003 with the expressed goal of lobbying for constitutional reform so that expressing one's sexual orientation would be acknowledged as a fundamental right within the Guyanese constitution. The group formed in response to political resistance to legal amendments that would have made it illegal to discriminate against someone because of their sexual orientation.[23] Though they currently have no full-time staff members and work primarily with consultants and project-based staff (Wills 2010, 14), they have the support of and provide support to a wide cross-section of other NGOs working in the area of sexuality rights (Wills ibid.).

I want to explore somewhat illustratively some of the ways in which SASOD has been engaging with what I am referring to as symbolic capture in the interest of sexual survivance since I see SASOD as already having begun a process that vernacularizes the spirit of the Yogyakarta Principles within the Caribbean. Whether through control of mobility, processes of myth making, truth-making, or self-surveillance, the state's desire as an expression of power is always to control those spaces and bodies that announce its vulnerability (Foucault 1995). SASOD challenges this by simultaneously inhabiting a number of different political terrains so that it is immediately inter/national. SASOD has located itself as an international organization that resides in Guyana. I state this to trouble the sense that "international" signifies the North; rather, I want to think of an international where global and local specificities have space to reside as co-constituting elements.

This layered identity is achieved through a number of different practices. Despite limited dedicated staff, SASOD is supported by a globally dispersed Caribbean queer community of almost 200 that remains connected through SASOD's online presence, which includes a Web site, blogs, discussion forums, and LISTSERV communications. These communications themselves educate on a range of political activities affecting LGBT/queer communities in the region and farther afield in the world.[24] This attention to activism, legislation, and policy elsewhere recalls Mohanty's articulation of a transnational praxis of activism, which allows us to envision a form of sexuality rights activism that resists "activism and agency in terms of discrete and disconnected cultures and nations (and) allows us to frame agency and resistance across the borders of nation and culture" (2003, 243).

SASOD calls the state into a relationship of accountability with international conventions and arenas. It was one of the first organizations in the region to articulate its political issues through the lens of the Yogyakarta Principles. Even before Yogyakarta was launched, SASOD began to invoke its reframing authority in order to challenge the legitimacy of a forum organized by the Ministry of Health, the National AIDS Programme Secretariat, and the Guyana Teachers' Union to debate the topic "Teachers who are homosexual/lesbian should not be allowed to teach."[25] In the realm of everyday activism, influencing the tone of a debate is, at times, more productive than attempting to stop it. To this end, SASOD's invoking the yet to be released Principles positioned the upcoming event within the purview of an international human rights discourse, prompting the participants to consider the deleterious effects of supporting discrimination in the context of public/private employment as a result of sexual orientation/gender identity (Principle Twelve).[26]

SASOD's symbolic capture readily shows the ways in which counter-discourses must for the purposes of intelligibility work simultaneously within and against the systems that it challenges. Therefore, it is important to also attend to the ministerial and sectoral work that they have accomplished, work that allows them to touch the functioning of the state in transformative and responsive ways to questions of sexual orientation and gender identity. Out of a two-year partnership between SASOD and the Ministry of Health came a training manual designed to enable health care providers to deliver optimal care to LGBT groups that had previously raised issues of discrimination within the system. Recognizing that this is the same Ministry that SASOD would have lobbied earlier for their participation in the above debate on queer teachers in the educational system, we also learn in the context of symbolic capture the importance tactically of moving in and out of an unpredictably contentious relationship with the arms of the state. Additionally, SASOD's challenges reflect the transformative aspect that results when the "we" of the everyday are called on to imagine queer subjects as having a right to everyday mythmaking spaces, such as pedagogical sites of encounter.

Seizing public domains for a range of queer expressions is a critical component of symbolic capture. For six of its seven years of existence, SASOD has held what is now an open, public gay and lesbian film festival. So, although SASOD began with a declared mission of constitutional change, it has developed an advocacy platform that makes LGBT subjects socially, culturally, and politically intelligible in Guyana—in ways that reflect a right to assemble peacefully (Yogyakarta Principle 20) and the right to participate in the cultural life of Guyana, indeed, to name queer culture as a subset of Guyanese culture (Yogyakarta Principle 26).

The festival might be seen as an example of how transnational circuits of cultural circulation produce a sense of queer cosmopolitanism. My use of queer cosmopolitanism should not suggest the homogenizing force of a "global gay" (Altman 2001). Instead, it is informed by Jon Binnie's careful discussion of the limitations of cosmopolitanism which conflates the idea with certain cities, elites, or practices of consumption (2004:126–127). Cosmopolitanism often connotes a kind of "worldliness and knowledge (i.e., knowing which brands to buy; Binnie, 128). Yet, I think that to articulate cosmopolitanism from and in relation to the global South, from the periphery, fractures this understanding and suggests a sense of "in-the-worldness" rather than the limited idea of "worldliness." Many of the films screened in Guyana might be seen in LGBT film festivals in Washington, DC, Portland, or San Francisco. However, more interestingly, and certainly not the case with the LGBT film festivals of Washington, DC, Portland, or San Francisco, SASOD's 2010 film festival was billed to reflect both the local and diasporic diversity present in Guyana—"14 nights, 39 films, 17 countries, 9 languages." The film festival not only carves space within the local social collective but also bridges a sense of an unstable belonging to a queer cultural circuit that moves beyond a United States–Caribbean flow. This international circuit certainly prompts some consideration of the place of LGBT film festivals in structuring a sense of a locally mediated connection with a set of epistemologies about queer identities.[27]

Earlier in this project, I pointed to the ways in which the ruse of modernity, through its many indicators and rankings, aimed to assess the "progressive" nature of the state based on how it treated "its" women and sexual minorities. This process takes a contradictory turn on issues of sexuality rights in the Caribbean. Caribbean states strongly articulate their identity in dialogue with the hauntings of colonial time, which prompts a discriminatory stance toward queer subjects. As such, contemporary rankings do not carry similarly seductive powers as has occurred in the race to gender-based ratification. So, while Guyana has signed on to the Resolution on Human Rights, Sexual Orientation, and Gender Identity (AG/RES. 2435 XXXVIII-O/08–June 2008), as SASOD notes in articulating the rationale for its work, there remains legislation that permits discrimination based on sexual orientation and gender identity, punitive sanctions against cross-dressing, and the need for gender-neutral rape laws that allow for a recognition of male rape.[28]

SASOD prompts a leveling of this uneven legislative landscape through a range of challenges that locate the everyday trauma of the absence of protective measures. SASOD's work has been a powerful challenge to impulses that aim to erase LGBT subjects from public/political landscapes. Its engagement with questions of equity and justice highlights the work and lives that communities forge in the midst of the region's variegated hostilities.[29]

SASOD's challenge to state legislation that punishes forms of cross-dressing has been one of the most frontal assaults on the process by which people get to count. In February 2009, seven male-to-female transgender Guyanese were rounded up in a crackdown. They were stripped, denied medical attention, detained over a weekend, and fined $7,500 (US$36) under Section 153(1)(xlvii) of the Summary Jurisdiction (Offences) Act, Chapter 8.02. Unrepresented, they appeared before Guyanese Chief Magistrate Melissa Robertson on February 9, 2009. Four of the seven men pleaded guilty to the charge of dressing in female's attire and damage to property, while the remaining three pleaded not guilty to the charge of dressing in female's attire and loitering.[30] The disciplining gaze of the state can be seen in the ways in which their court appearance was recounted within the public media.

They were ridiculed by the magistrate from the bench, lectured that they were men, not women, admonished that they were confused, and instructed to go to church and give their lives to Jesus Christ.[31] The details regarding their evening's dress also smacked of state sanctioned voyeurism, suggested by the detail regarding their apparel:

> Bess stated that on the day in question they were not wearing female clothing but "unisex" clothing. He said that on the day in question he was wearing jeans, a designer top and a hooded overcoat.
>
> Peters stated that he was wearing a "skirt with a designer top."
>
> Fraser then stated that he was wearing a "black dress with green designs."

Persaud stated that he was wearing a "short skirt with a red designer top."

Each account of their dress becomes a declaration designed to position their dress as an act of undressing, remembering that how the state calls forth the trans body to "self-identify" performs all sorts of epistemic violations. This then becomes a state of undress that requires them to account for, if not take off, their sexual performance identity through a public practice of disavowal.

The 2009 cases generated considerable publicity, and there were many domestic and international appeals to the Guyanese Government to remove the law. After these went unheeded, the constitutional motion was filed on February 19, 2010.[32] The motion was filed with the support of lawyers in Guyana, St. Lucia, the University of the West Indies Rights Advocacy Project (U-RAP) from the Cave Hill, Barbados, campus, and SASOD. In a press release issued on the day that the motion was filed, litigant Seon Clarke, also known as Falatama, lamented: *"It was one of the most humiliating experiences of my life. I felt like I was less than human."* These cases mark the state's capacity to forcefully delineate the line of in/humanity. The irony of SASOD's engagement with scripts of in/humanity is that a degree of its success is predicated on a form of activism that is only intelligible because it draws on the same logic, terms, and discourses that desire to silence it. It is a queer positionality which then makes it possible to resignify, that is to say, to name these terms otherwise.

As a medium, the Internet is now firmly entrenched as a site of queer subject formation.[33] However, what of queer subjects in areas where Internet access is uneven at best, and print media continue to hold significant national presence? In this context, the public debates that followed within the print media (with online presence) are an important discursive site to examine. On one hand, the online discussions facilitate a transnational presence and intimacy to the debates.[34] On the other, for queer subjects outside of the community, the public debates that occur between SASOD supporters and detractors announce to local queer subjects a disarticulated connection and distant intelligibility, that is to say—"someone like."[35] The media discussions in this sense are expansively transnational and profoundly local.

This simultaneity of the transnational and the local is illustrative of the importance of embodied spatialities to queer organizing. Mindful that each of the articulations in the public debate is not similarly situated, i.e., speakers matter, as does their location, I still think it is useful to understand this transnational conversation as one where the politics and place of the "local" are conveyed through and influenced by embodied migratory shifts and diasporic expansions. The Caribbean diaspora is so expansive that whether an individual queer subject travels or not in no way fully captures the flows of cultural, ideological, or activist interchange. The idea of embodied spatialities attempts to capture how space is influenced by what kinds of movements are possible—which ideas about identity resonate and circulate, how does such circulation occur, how easily do bodies move, and what kind of transnational organizing occurs around a given idea of justice, rights, or entitlement.

This approach to the "local" enhances our understanding of the kinds of activism at work within societies with large diasporic populations. Rendering the local as that which travels in multiple directions through persons, we are further able to address another layer of complexity to equity and justice claims that include and exceed South–North arrangements, to address South-South, as well as diasporic South-intra-North activism and deployment of resources. By embodied spatialities, the local is expanded to incorporate multiple spheres of travel, activity, and circulation of issues.

SASOD's legal challenges contest the politics and conditions of emergence of sexual minorities in the Caribbean and, by so doing, not only encapsulates the constraints that mitigate against sexual minorities appearing in the public sphere but pushes visibility beyond the spectacle and the carnivalesque to the recognition that it is equally as important to access the places, spaces, and practices that facilitate personhood. Here, we might be mindful of Berlant's observation that "citizenship is a status whose definitions are always in process. It is continually being produced out of a political, rhetorical, and economic struggle over who will count as "the people" and how social membership will be measured and valued."

The motion filed to repeal the laws on cross-dressing in Guyana opens a conversation about who gets to count and how this will be measured with a political carefulness that far exceeds the descriptive butchery of the label homophobic. It captures the ways that "travel" within the context of a queer discourse incorporates but extends beyond questions of physical mobility. The ongoing debates and discourses on sexuality shed light on what it means for some subjects to be peripheral within the periphery, while simultaneously pointing to moments of possibility within the region's variegated hostilities. This project has centered the politics of queer space/place as an engaged struggle of seizure that aims to undo the rhetoric of tolerance while offering a more complicated engagement with the contradictions of entitlement and rights.

CODA

It was originally my hope to have titled this book *Cartographies of Being: Citizenship, the Body, and Feminist Advocacy in the Anglophone Caribbean*. In the world of keywords and indexing, "cartographies" was vetoed as signaling a market of perplexed geographers. I disagree. I don't think they would have been perplexed.

However, having conceded to larger imperatives, I want to return to why the categories of cartography, body, citizenship, and advocacy were important to this project. What I have been attempting to do in *Feminist Advocacy* is to map the ways in which ontological processes come into play through policy, law, and various circuits of travel. How these artifacts travel and mark themselves on differently situated bodies, therefore, has been a recurring theme throughout this text. How their meaning differs as they touch different geographies has also been a critical plane of interrogation. It is in this context that future iterations of this project point to a queer geography as a way of understanding the ways in which queer subjectification is constrained by questions of place but equally allow for a mapping of the trajectories through which queer subjects mark their embodiment on the surrounding landscape.

Analytically, the category of "time" appeared in this project with a great deal more potency than I would have anticipated when I started this work. It appeared as I considered the place of institutional histories in contemporary implementation of gender mainstreaming. Once again, as I encouraged us to consider the relevance of historicizing gender-as-genealogy within contemporary gender mainstreaming approaches, and, of course, finally, we examined the effects of colonial time on modern queer subjectivities. The implied connection here does indeed bring us to the ways in which citizenship is constantly in process, weaving its way in and through time so that different subjects are able to contest the ways in which earlier articulations have foreclosed on the possibilities of belonging.

I have toward the end of this text centered a queer subjectivity as a particularly liberatory location to some of the vexing strategies of un/belonging. However, this act of noticing queer subjectivity has only been for the purposes of experimentally exploding the "we" of the everyday. An explosion of the kind of heteronormativity that finds pleasure, to draw on Lauren Berlant's work, as it "uses cruel and mundane strategies both to promote shame for non-normative populations and to deny them state, federal, and juridical supports because they are deemed morally incompetent to their own citizenship" (1997, 19).

It is my hope that in some way *Feminist Advocacy* contributes to a form of advocacy that understands the nuances of what Braidotti has pointed to in a different context as blurring the boundaries without burning the bridges. For as much as we may wish to burn bridges, I believe that the residual effects of oppression must sit with us for purposes of remembrance even as we reach for a more just society. Additionally, I hope that I have been adequately mindful of the transnational lesson that what we do when we encounter borders affects the effectiveness of our politics. Therefore, have I listened, trampled, or sat appropriately with the borders of this project? I leave you to decide and to continue the conversation with me.

Notes

NOTES TO THE PREFACE

1. My use of "developmentalism" aims to convey the economic determinism that informed development policies and approaches applied to the global South. Tony Smith's "Requiem or New Agenda for Third World Studies?" (1985) discusses the growth of developmentalism within development policies. See for example, Arturo Escobar's *Encountering Development: The Making and Unmaking of the Third World* (1995) for a critique of a version of "developmentalism" that connotes over-consumption, unwavering faith in the market, and fictions of the "South."
2. Historical critiques were central to the dependency and de-linking approaches of the 1960s and 1970s. These connections were critical to understanding the "underdevelopment" of the "Third World" as historically structured relations of inequity rather than a failure of economic tools and measures at the national level.
3. The term "national machinery" was first introduced in 1975 at the First UN World Conference for Women in Mexico. The Platform that came out of this conference was the first UN instrument to call for a range of support mechanisms, policies, and personnel within the state dedicated to the achievement of women's advancement. This call resulted in the establishment of women's ministries, desks, and bureaus with varied levels of success in achieving their goal. (UN Development Program 2005; UNDP).
4. Hester Eisenstein's *Inside Agitators: Australian Femocrats and the State* (1996) provides a comprehensive discussion of the dilemmas of state-based feminism while mapping, nonetheless, the potential of feminist bureaucrats to initiate change from within the national machinery.

NOTES TO CHAPTER 1

1. My community is the Anglophone Caribbean which refers to the independent English-speaking nation-states of the region; these include Antigua and Barbuda, The Bahamas, Barbados, Dominica, Grenada, Jamaica, St. Kitts and Nevis, St. Lucia, St. Vincent and the Grenadines, Trinidad and Tobago, as well as Belize and Guyana on the continents of Central and South America, respectively. I'm also including in this constellation the British overseas territories of Anguilla, British Virgin Islands, Cayman Islands, Montserrat, and Turks and Caicos Islands. I am guided here by the commonality of British colonialism, economic vulnerabilities of scale or weak local industries, and similar parliamentary systems of government. These territories are also

(differently) grouped within the Caribbean Community (CARICOM)—the Community aims to work collectively to improve the standard of living and work in the region, as well as "coordinated and sustained" economic development. The British Virgin Islands, Turks and Caicos, Anguilla, Cayman Islands, and Bermuda are presently Associate Members rather than full members of CARICOM, as are the other territories identified above. The region, however, is probably one of the most ethnically diverse regions in the world, with people of Amerindian, South and East Asian, African, European, and Middle Eastern descent. See Rhoda Reddock's *Ethnic Minorities in Caribbean Society* (1996) for an additional discussion of this diversity.

2. Shirin Rai's and Jacqui Alexander's work inform my thinking on state formations in *Feminist Advocacy*. Rai offers a number of critical insights regarding state formations in the global South. In *The Gender Politics of Development*, Rai emphasizes the fractured and untidy nature of the state. Women, I suggest, experience this lack of unity with a measure of temporariness and arbitrariness; in other words, there is no guarantee that regulations, goals, and policies cohere from one government agency to the other. Rai also notes that, qualitatively, women experience the regulatory power of the state to greater degrees than spaces governed by European liberal systems of government (2008, 58). This theme is one that emphasizes Jacqui True's observation that women cannot afford to not interact with the state in the achievement of their goals. I also employ Jacqui Alexander's use of the term "state-manager" in my discussions of the state. I use this term for both bureaucrats and politicians alike. However, when it is important for me to reference specific politicians, I do so. What I draw from Alexander's usage is the question of intentionality. The use of the term "state-manager" reminds us that the *state* never does anything without witting agents acting in its name (Alexander, 1994). Situating these agents or managers also has analytical merit as it better positions us to challenge the disembodied veneer of state rationality that sidelines the relevance of women's concerns. My caveat here, however, is that at no point should we see state power as coterminous with given actors and decision makers; while these agents do have the power to influence politics, they also operate within discursive frames that exceed their individual location. However, even as the use of the term "state-manager" prompts us to think about the embodiment of state action, its limitation is that it in no way helps us to identify the politics of the actor; for that, we need to trace the actor's exercise of power.

3. See, for example, "Institutional Mechanisms for the Advancement of Women" within the Beijing Platform (UN 1996).

4. I'm using "additive" to infer the more popular usage of "add women and stir," a reference that critiques the rhetorical integration of women into areas formerly dominated by men. This term is used somewhat pejoratively by feminists to critique the ways additive discussions omit a commensurate analysis of the systemic conditions that produce and perpetuate gender asymmetries.

5. The Westminster system of government is a democratic, multi-party parliamentary system modeled on the conventions of the U.K.'s Parliament. It is characterized by three separate arms of government (executive, legislative, and judiciary); the head of the executive is in practice the head of the largest elected party, thereby making the Westminster model a "first-past-the-post" or "winner-takes-all" approach. The head of the majority often wears the title of Prime Minister. This majority is an imperative if bills are to be passed; failure to pass bills, such as the budget, could result in the dissolution of Parliament. The need for a majority sometimes stifles any possibility

for minority or oppositional practices within the party. Achieving gender equity in this model is made more difficult if gender issues are allocated to a "soft" ministry or lowly positioned minister. See Payne and Sutton (1993) for an extended discussion on the application of the Westminster system in the Caribbean.
6. See Jo Beall's "'Trickle Down or Rising Tide?' Lessons on Mainstreaming Gender Policy from Colombia and South Africa" (1998) for a comparative discussion on gender mainstreaming.
7. Here, I am referring to the Caribbean's participation in the wider grouping of the Commonwealth of Nations, established in 1971 under the Singapore Declaration. The Commonwealth Secretariat's Gender Unit has also integrated gender mainstreaming into its own operational processes and programming, and has taken the lead globally in the thinking on gender budgets as a fiscal mainstreaming instrument.
8. See also Alvarez's (2000) discussion of the UN bureaucratic logic at work in such transnational partnerships, which brings with it inherent hierarchical demands that are easily able to overpower existing modes of operation.
9. By the term "clientilism," I am referring to the forms of political patronage (e.g., jobs, improved infrastructure, and money) that politicians extend to members of various constituencies in return for votes.
10. I am working here with Linda Alcoff's use of positionality as that which allows us to acknowledge how the concept "woman" may be identified by a set of attributes and networks of socio-economic relations without placing women in a fixed relationship with these attributes. Alcoff notes, "If we combine the concept of identity politics with a conception of the subject as positionality, we can conceive of the subject as non-essentialized and emergent from a historical experience and yet retain our political ability to take gender as an important point of departure. Thus, we can say at one and the same time that gender is not natural, biological, universal, ahistorical, or essential, and yet still claim that gender is relevant because we are taking gender as a position from which to act politically" (1988, 427).
11. For a more in-depth discussion of these debates, see Carol Bacchi and Joan Eveline's "What Are We Mainstreaming When We Mainstream Gender?" (2005). Their discussion of the Netherlands' resistance to a sex/gender distinction in favor of "gender as a political process" is an important example of what is possible when feminism is not divested from state processes of mainstreaming. Also, see Baden and Goetz (1998).
12. http://web.worldbank.org/WBSITE/EXTERNAL/TOPICS/EXTGENDER/0,,contentMDK:20193040~pagePK:210058~piPK:210062~theSitePK:336868,00.html (accessed August 14, 2008).
13. Baden and Goetz (1998) and El-Bushra (2000) all call into question the statistical improbability and reductionist aspects of UN slogans such as "Women account for two-thirds of all working hours, receive only one-tenth of the world's income and own less than one percent of world property" (UN 1980). Their aim is not to question women's disadvantageous access to the world's resources. However, they do share some concern about the computation of statistical evidence as well as the ways in which such sound bites are abstracted from the complexity of their context.

NOTES TO CHAPTER 2

1. This economic determinism emerged more formally in the 1970s and 1980s within the "WID" approach that drew heavily on Ester Boserup's seminal

work on agricultural inequity in Africa. WID highlighted women's and men's differential experience of development, was characterized by the establishment of desks, programs, or ministries within national machinery to facilitate the full integration of women's concerns into national policy and planning agendas and trained women to maximize their participation in the formal economy. During this period, feminists struggled as well to make state policy more responsive to the needs and concerns of women as agents of development (Moser 1993).
2. Loosely translated as "boys will be boys."
3. The Apprenticeship period refers to that four year transitional period (1834–1838) in which British Caribbean slaves were required to function as waged apprentices prior to complete emancipation in 1838.
4. Rhoda Reddock uses Maria Mies' term "housewifization" to provide a very useful discussion of colonial attempts to deploy female labor (physical and sexual) into the domestic sphere during the period of wage-driven labor.
5. The Amelioration period was characterized by colonial attempts to "improve" the conditions of slavery. These policy shifts were instrumental in nature to stave off abolitionist pressures and piecemeal in implementation (Beckles 1989; Morgan 2008).
6. The term "post-colonial" reflects both anti-colonial sentiments of nationalist movements as well as the colonial continuities in national state formations (Loomba, 1998).
7. Brereton, for example, asserts, "Production took priority over reproduction and gender distinctions mattered relatively little in the day to day operation of the slave plantation" (1988, 131). This type of argument works on the assumption that, as renewable chattel, slave women's reproductive capacity did not figure centrally in the economic decisions made by colonial planters.
8. This is to be compared with the 2,900 hours that a factory hand in England would have worked during the 1830s (Higman 1995, 188).
9. See Michael Craton's discussion of the colonial rationality that dictated an emphasis on breeding rather than buying new slaves (1978, 19).
10. Trinidad and Tobago's independence would come later in 1962; however, the growing political sensibility of this period was a pivotal precursor to independence.
11. The Negro Welfare Cultural and Social Association (NWCSA), formed in 1934 in response to the Abyssinian War, was also instrumental in mobilizing a wider base of protest in Trinidad (Singh 1987, 62).
12. The findings of the Commission received a skeptical response. The findings were seen as primarily cosmetic and in the words of calypsonian *Atilla,* "A peculiar thing about the commission/And their ninety-two pages of dissertation/Is that there is no talk of exploitation/Of the worker or his condition." It is not surprising, therefore, that the members of the commission advocated that the existing labor unions be neutralized (Rohlehr 1990, 209).
13. The WIRC is also known as the Moyne Commission, named after Lord Moyne (Walter Guiness, Secretary of State for the Colonies—1941–1942), who headed the Commission. I will use both names interchangeably.
14. Mr. Till had just argued that the machinery depreciated proportionally to his workers' wage due to poor handling because the workers did not have a "machine outlook" (*Trinidad Guardian* January 4, 1939, 1).
15. Rhoda Reddock, citing Ron Ramdeen, notes that the events of 1937 were narrated with language such as "a band of men" or "several hundred" and again "three busloads and one lorry load of men" (1994, 157), this despite the fact that many of the agitators, if not instigators, of the riots were women (ibid.).

16. I use the juxtaposition of *eye/I* as a way of "languaging" the intimacy between knowledge production and subject formation (Rowley 2003, 26). However, the defining aspect of the relationship between the "eye" and the "I" is the "power to name" that society confers on certain vantage points (e.g., the "eye" of power brokers, such as priests, politicians, and social workers) inevitably named (the subject "I") from their vantage point and informed by their biases.
17. It is not accidental, therefore, as we will examine later, that the first individual to head the Colonial Welfare and Development Office in the region was renowned British sociologist T.S. Simey.
18. Simey's commitment to an anthropologically informed social order was evident in his harsh critique of the WIRC's lack of foresight in not centering sociology or anthropology in their deliberations for a West Indian university. He maintained that changes in the social patterns of West Indians required both integrated planning and scientific involvement (1946, 233).
19. Structuralist assumptions aimed to create ideal type yardsticks, that is to say, prescriptive models, which, if attained, would result in greater levels of social equilibrium. The sociological prescription of nuclearity, therefore, ought not to suggest that this was necessarily so on the ground in England or the United States, but rather, that it was the ideal around which policy and gender role expectations would be established.
20. I use the phrase "dismembered women" to highlight the recognition of women only, and primarily, in their capacity as mothers, to the exclusion of other intersecting subjectivities. It also reflects the paradox of the "maternal" being where the location is assigned a deceptively high social value, while all other dimensions of the life remain in a state of institutional crisis.
21. Among its various amendments, this Act has been amended in 1980 and 1996 and The Public Assistance Regulations of 1997 and 1998.
22. The Central Statistical Office defines this term based on the person "accepted or recognized as" by other members of the household responsible for the overall functioning of the household. This is still conceptually inadequate as most research instruments require their interviewees to identify the head of household. However, the prevailing gender ideology is such that it often delimits the extent to which women are willing to self-identify if there is a male partner present.
23. For a male head of household to be considered for public assistance, he has to be certified medically unfit or incapacitated. This in itself is problematic since, in the face of steady economic decline and shrinkage of the job market, the individual will receive no assistance if unemployed. A woman, therefore, cannot approach the state on behalf of herself or her children in this context because of the presence of an "able-bodied" male in the home who is designated first and foremost as "worker" and "head."
24. Interviews with social workers were conducted on site at a welfare office in Port of Spain, Trinidad.
25. Lazarus-Black uses the term "rites of domination" to refer to the events and processes that are enacted within court procedures that result in the maintenance rather than the disruption of class and gender hierarchies (2003, 4).
26. Lazarus-Black also speaks to the need for excessive, if not dramatic, production of "proof" required by women in order to lend credibility to their voice within the courts. She observes that this can extend to lawyers, bringing children to court for the magistrate to determine whether the child's physical appearance does not corroborate the woman's claims of paternal responsibility (ibid., 18).

27. Announcing an arrival may not necessarily mean a phone call. More often than not it is implied repeatability so that the recipients often said, "And they normally come on . . . but I didn't know they were coming that day," thereby suggesting a routine, but with the freedom for this routine to be broken. It is this element of the unexpected that succeeds in producing self-surveillance.
28. Interview conducted at home.

NOTES TO CHAPTER 3

1. The primary data for my overview of institutional mainstreaming mechanisms in the Anglophone Caribbean are used with permission and draws from research done by the author as commissioned by the UNIFEM, in preparation for the Fourth Ministerial Conference for Women in the Caribbean (February 12–13, 2004). I electronically mailed my questionnaire to twenty countries with a 50 percent return rate. This study included the territories of Anguilla, Barbados, Belize, British Virgin Islands, Cuba, Dominica, Jamaica, St. Kitts and Nevis, St. Lucia, and St. Vincent and the Grenadines. These data are reproduced with permission. All other interviews were conducted in Trinidad and Tobago during the period of 2000–2002 for my doctoral dissertation.
2. See Rhoda Reddock's "Women's Organizations and Movements in the Commonwealth Caribbean: The Response to Global Economic Crisis in the 1980s" (1998) for a discussion of some of the region's early feminist legislative milestones.
3. Hester Eisenstein traces the first use of the term "state feminism" to Norwegian social scientist Helga Hernes (1987) who used it to refer to the establishment of political machinery and policies aimed at addressing women's concerns within state apparatus (1996, 217).
4. Veitch identifies a number of important legislative changes, such as child care tax credits and legal protections for those opting to use part-time employment, but she notes simultaneously that many of these pieces of legislation do little to open new possibilities for women (2005, 605).
5. Interestingly, the two cases above both come from a 2005 comparative conversation on gender mainstreaming in the *International Feminist Journal of Politics*. While mainstreaming practices in the South are referenced in the issue, the lack of voices of mainstreaming practitioners *from* the global South remains the issue's primary weakness.
6. In her four-donor agency study (Canadian International Development Agency, Norwegian Agency for Development Cooperation, UNDP, and World Bank [WB]), Jahan argued that effective gender mainstreaming processes would require a two-pronged approach, namely, one which would initiate *institutional and operational* changes to daily activities (1995).
7. This chapter references the reprint of Goetz's piece that focuses on only five of the countries originally studied (2007). It is important to note that her initial study included Bangladesh, Chile, Jamaica, Morocco, Vietnam, Mali, and Uganda. The latter two were omitted from the 2007 reprint, however. In keeping with the integrity of the 2007 reprint, I will refer only to the five countries cited there. Interviews for Goetz's study were conducted during the period 1994–1995.
8. See http://www.un.org/womenwatch/osagi/gmfpstudy.htm for an assessment of gender focal point's functioning within the UN system (accessed March 22, 2010).

9. This project occurred and was written in 2004; due to a typographical error, the published document has the issue date as 2003. For citation consistency, I have kept the date on the document. However, the content will refer to materials discussed and gathered in 2004.
10. These included agencies such as UNECLAC, Canadian International Development Agency (CIDA), Pan-American Health Organization (PAHO), Organization of American States (OAS), UNIFEM, Caribbean Association for Feminist Research and Action (CAFRA), and Commonwealth Secretariat. See Rowley (2003) for a country-specific overview.
11. In Trinidad and Tobago, this nationalized desire for development is constructed most comprehensively in the prime minister's *Vision 2020*, a comprehensive ministerial restructuring toward Trinidad and Tobago's achievement of developed country status by the year 2020.
12. Unless a country has signed the optional protocol of the CEDAW Convention, there are no sanctions or inquiries that can be applied to nation-states who do not implement or enforce the terms of the CEDAW Convention. The optional protocol allows individuals or groups to place written complaints before CEDAW for any violations of the Convention. No country in the CARICOM region has signed the optional protocol. All have ratified CEDAW.
13. Despite the fact that Afghanistan and the United States have both not signed the CEDAW convention, the representation of the veiled woman and the protection of this brown body legitimizes a number of encroachments against the state to which she belongs. Liberating the brown female body has become a common trope that is used to validate increasing military offensive, "civilizing mission" against the non-Western nation-state.
14. Upon Independence, the two main political parties in Trinidad and Tobago were the People's National Movement and the Democratic Labour Party. In 1976, two new political parties emerged: the United Labour Front, led by Mr. Basdeo Panday, and the Democratic Action Congress, led by A.N.R. Robinson. The two latter parties merged in 1981 to form the National Alliance for Reconstruction. In 1988, there was a further split back into their organic parts with one-half keeping the acronym NAR and the other coining the name Club 88, which eventually became the United National Congress. With the exception of the Democratic Labour Party, each of these emergent parties would have been given the opportunity to hold the reins of power by the voting public (Ghany 1988, 183).
15. I deliberately demarcate livelihoods, well being, and capabilities to refer not only to economic empowerment and enhancement but also to incorporate "quality of life issues" and the expansion of individuals to exist with a sense of entitlement and dignity when claiming these entitlements.
16. Sonja Harris (2003, 182) has observed that the women's movement has not organized a sufficiently coherent response to the marginalization debate. This is not quite accurate; there have been a number of responses by both male and female feminists in the region (see Lindsay 2003; Barriteau 2003; Figueroa 2003). The failure, however, is to be found in the modes of dissemination, which have yet to be as populist or as far reaching as Miller's circulation of male marginalization.
17. The IDB Non-Traditional Skills Training Program (TC-94-04-37-1) was a regional project that should have been implemented and managed by a number of technical institutes within the region. Trinidad was, therefore, unique because it was the only territory in which a Gender Affairs Division was the implementing site of the Non-Traditional Training Program.

18. Equally as important to where the bureau is located is the number of times it gets volleyed from ministry to ministry, thereby compromising the consistency of its work program and mandate (Harris 2003).
19. Women Working for Social Progress, the Network for the Advancement of Women, the Caribbean Association for Feminist Action and Research, and more recently, Advocates for Safe Parenthood and Reproductive Equity have all consistently and continually challenged the state's will to constrain women to traditional state-sanctioned gender roles.
20. http://www.finance.gov.tt/documents/publications/pub7.pdf (accessed March 29, 2010).
21. http://sta.uwi.edu/cgds/genderPolicyLetter.asp (accessed March 29, 2010).

NOTES TO CHAPTER 4

1. See the report "Electing to Rape: Sexual Terror in Mugabe's Zimbabwe" at http://www.aids-freeworld.org/images/stories/Zimbabwe/zim%20grid%20screenversionfinal.pdf (accessed May 17, 2010).
2. See, for example, Susan Bordo's "The Body and the Reproduction of Femininity" (1997) where she analyses the specific power differentials that work to discipline and normalize how the female body is implicated in the reproduction of gender norms.
3. The ICPD, a twenty-year Program of Action on population and development related issues, was adopted by 179 countries in 1994, Cairo, Egypt.
4. http://www.who.int/reproductive-health/unsafe_abortion/map.html (accessed November 25, 2008).
5. http://www.who.int/making_pregnancy_safer/topics/maternal_mortality/en/index.html (accessed November 7, 2008).
6. Africa (Egypt, Nigeria, and Uganda), Asia (Bangladesh, Pakistan, and the Philippines), and Latin America and the Caribbean (Brazil, Chile, Colombia, Dominican Republic, Guatemala, Mexico, and Peru). Singh further combined these data with additional data from five countries in sub-Saharan Africa (Burkina Faso, Ghana, Kenya, Nigeria, and South Africa).
7. The Millennium Development Goals (MDGs):
    1. Eradicate extreme poverty and hunger through halving the percentage of people who live on less than a dollar a day, as well as the percentage of people who suffer from hunger.
    2. Achieve universal primary education by ensuring that all boys and girls complete a full course of primary schooling.
    3. Promote gender equality and empower women by eliminating gender disparity in primary and secondary education, preferably by 2005 and at all levels by 2015.
    4. Reduce child mortality by reducing the mortality rate of children below the age of five by two-thirds.
    5. Improve maternal health by reducing the maternal mortality rate by three quarters.
    6. Combat HIV/AIDS, malaria, and other diseases.
    7. Ensure environmental sustainability by integrating the principles of sustainable development in country policies and programs. Halve the population of people who do not have access to potable water and improve the lives of at least 100 million slum dwellers by 2020.
    8. Develop a global partnership for development through the formulation of an open trade and financial system that is

rule-based, predictable, and non-discriminatory. Include a commitment to good governance, development, and poverty reduction—nationally and internationally. Address the least developed countries' special needs. This includes tariff and quota-free access for their exports; enhanced debt relief for heavily indebted poor countries; cancellation of official bilateral debt; and more generous official development assistance for countries committed to poverty reduction. Address the special needs of landlocked and small island developing states. Deal comprehensively with developing countries' debt problems through national and international measures to make debt sustainable in the long term. In cooperation with the developing countries, develop decent and productive work for youth. In cooperation with pharmaceutical companies, provide access to affordable essential drugs in developing countries. In cooperation with the private sector, make available the benefits of new technologies—especially information and communications technologies.

8. My use of "globalised developmentalism" should not be confused with a dependency school's use of developmentalism, which Immanuel Wallerstein notes became the "code word for the belief that it was possible for the countries of the South to 'develop' themselves, as opposed to 'being developed' by the North" (2005, 1264).

9. Note, for example, the Caribbean Community Heads of Government meeting held in Washington, D.C., June 19–21, 2007. The meeting was strategically hosted in the United States in order to resuscitate a perceived waning of U.S. interest in bi-lateral relations between the United States and the CARICOM region. With this political intent in mind, a quick perusal of the conference's Web site (http://www.conferenceonthecaribbean.org/Background/tabid/57/Default.aspx) reveals an interpretation that focuses primarily on economic security-related dimensions of the MDGs.

10. Note that the indicators for MDG Five (improving maternal health) are maternal mortality and increase in percentage of births attended by skilled health care providers.

11. "A Caribbean Community For All." Compton Bourne, President of the Caribbean Development Bank, June 25, 2003. Fifth Lecture in the Distinguished Lecture Series Celebrating the Thirtieth Anniversary of the Caribbean Community.

12. In this paper, I hold reproductive health and sexual well-being in tandem with each other in order to expand the discussion on reproductive equity beyond a procreative, health-related matter. Sexual well-being, therefore, incorporates non-heterosexual identities as well as the importance of non-procreative sex.

13. See Arabella Fraser's insightful critique of MDGs in relation to maternal mortality, which places this global concern in a historical context that dates back to the Safe Motherhood Conference of 1987 aimed at reducing maternal mortality by 50 percent by the end of the decade. Fraser reminds us that high maternal mortality rates more often than not indicate a weak public health care system (2005). This observation is important because it places the MDGs in the context of global and national systemic conditions: a feature that receives hardly enough attention since many of the MDGs require economic integrity within the nation state. Yet, in their formulation, the goals do not consistently (within each goal) discount for the demands of globalization, nor apportion social responsibility as a result of these inequities.

214 *Notes*

14. Caribbean Development Bank (2005, 30).
15. I've used this term to highlight the recognition of women only, and primarily, in their capacity as mothers, to the exclusion of other intersecting subjectivities. I have in the past also used this term to reflect the paradox of this "maternal" being in which the location is assigned a deceptively high social value while all other dimensions of her life remain in a state of institutional crisis (2003).
16. See "Millennium Development Goals and Sexual and Reproductive Health: Briefing Cards." Family Care International (2005).
17. Statement by Kofi Annan, United Nations Secretary General to the 5th Asian and Pacific Conference on Population Control (UNESCAP Bangkok 2002).
18. See "Millennium Development Goals and Sexual and Reproductive Health: Briefing Cards." Family Care International (2005, 3).
19. See Dorothy Roberts's discussion of how critical family planning and population control were to Trinidad and Tobago's development trajectory in the 1960s. She further examines the rationale of the demography in that era by suggesting that women would be more favorably inclined toward contraceptive use if they understood the national problems (2003).
20. See PAHO MDGs at http://www.paho.org/english/mdg/cpo_meta6.asp (accessed March 12, 2010).
21. See Roberts (2003).
22. Though Guyana's legislation occurred in 1995, the lobby and consultation for the greater part occurred prior to 1994.
23. In Trinidad and Tobago, termination of pregnancy is prohibited under the Offences Against the Person Act, Sections 56 and 57. Section 56 states that "Every woman, being with child, who with intent to procure her own miscarriage, unlawfully administers to herself any poison or noxious thing, or unlawfully uses any instrument or other means whatsoever, with the like intent, and any person who with intent to procure the miscarriage of any woman, whether she is or is not with child, unlawfully administers to her or causes to be taken by her any poison or other noxious thing or unlawfully uses any instrument or other means whatsoever with the like intent, is liable to imprisonment for four years." This prohibition was modified in 1938 in the English case of the Queen versus Bourne, in which an obstetrician who had performed an abortion on a fourteen-year-old who had been raped was cleared of the charge in favor of the defense that "abortion was not illegal if done in good faith to preserve the life of the mother and to ensure she did not become a mental wreck" (Charles 2000).
24. In some ways, this instrumental approach becomes a gendered exampled of Baugh's recurring theme of "who trouble me trouble you." In other words, everyone benefits economically when the leader alleviates his own accounting woes through the management of women's reproductive options.
25. See Antigua and Barbuda's Offences Against the Persons Act, Cap 300: Sect. 56: "Administering Drugs, or using instruments, to procure abortion."
26. See Barbados' Medical Termination of Pregnancy Act, Cap. 44A Section 4, which allows for terminations beyond a twelve-week period if the medical practitioner is of the opinion that the pregnancy presents a risk to the mother's life or proves to be injurious to her physical or mental health or that of her child. Further, Barbados also makes allowances for terminations on the basis of socio-economic hardship, both actual or foreseeable.
27. See Guyana Medical Termination of Pregnancy Act 1995 (Act No. 7 of 1995). In the Guyanese context, terminations that are performed between eight and no more than twelve weeks must be done by an authorized medical practitioner, in an approved institution, and only in the context of rape, incest,

injury to mother's and or child's mental or physical health, HIV status and where there is evidence that the pregnancy occurred despite the use of a recognized contraceptive method. Terminations before the eighth week should be performed by an authorized medical practitioner, need not adhere to the above stipulations, nor in an authorized institution. Terminations that extend beyond the twelfth week, but not beyond sixteen weeks, should only be done in the context of grave permanent injury to child or mother. With regard to children, the act states that practitioners should encourage the child to inform the parents, but neither child nor practitioner is obliged to inform parents.

28. I've opted to use the term "termination" rather than the more popularly resonant "abortion" in keeping with the legislative and advocacy language used in the Caribbean.
29. My use of a relay is purely metaphoric, which is possible only from hindsight. It should not suggest a smooth, prescient handing over of a reproductive rights baton. Nunes's (1995) work suggests, however, that subsequent reproductive rights advocates did build on prior openings in their activism with the Pro Reform Group's bringing the final legislation baton home.
30. This point emerges out of a conversation with Tracy Robinson, Faculty of Law, The University of the West Indies, Cave Hill Campus.
31. I want to thank Tracy Robinson for helping me clarify this point.
32. Dame Billie subsequently served as chair of the NGO Planning Committee for Cairo and as President of the International Planned Parenthood Federation/Western Hemisphere Region.
33. Then President of Barbados Family Planning Association Mr. George Griffith argued that Barbados was ready for the decriminalization of abortion out of a concern for both the high level of maternal mortality, particularly in the rural areas of Barbados, and the subsequent strain on state resources from maternal morbidity (interview with author June 2003).
34. The Cabinet called for a non-partisan committee comprising of the Barbados Workers Union, the National Union of Public Workers, the Barbados Union of Teachers, the Barbados Association of Medical Practitioners, the Barbados Secondary Teachers Union, the Bar Association, the Association of Social Workers, the National Organization of Women, the Barbados Ministerial Association, the Ministry of Legal Affairs, the Barbados Ministerial Association, and the Manager of the Barbados Family Planning Association.
35. St. Lucia has a population of approximately 160,000, 80 percent of which is Catholic.

NOTES TO CHAPTER 5

1. The research for this discussion was done by the author in 2005 on behalf of CASH. Interviews were conducted at the following stations: Holetown, Central Division, Oistins/Southern Division, District A, District E, and the Training School, with interviewees taken from all ranks. I am grateful to the RBPF Senior Command team's courageous willingness to confront issues of gender equity in their organization and placing themselves under such scrutiny, to members of CASH for their guidance, and to the UNIFEM Regional Office for their financial support of CASH's work.
2. At the point of the study, there was only one female member of the Senior Command Team. I promised all interviewees anonymity whether they requested it or not. In an effort to ensure this, my citation will not identify whether senior members are male or female. I specify gender only at the lower levels.

216 Notes

3. Note, for example, in Table 5.2 the disparity in male/female pass rates in the one year that the number of male applicants dropped significantly.
4. The interview process consists of eight exams conducted over a period of four months.
5. Within the RBPF's human resources materials, the term "wastage" features prominently in discussions of personnel use of time. It refers to any loss of personnel as a result of a wide range of issues such as medical grounds, personal days, retirement, death, etc.
6. My distinction between hegemonic and normative here aims to capture the ways in which certain gender identities may perform in hegemonic ways organizationally, while not necessarily functioning as normative in the wider society.
7. To date, the organization has produced a 68-page report entitled "Sexual Harassment and the Law in Barbados" and a 66-page research report entitled "Charting Gender Equity in the Workplace, With a Special Emphasis on Sexual Harassment: A Case Study of the Royal Barbados Police Force." They held two public fora in October 2003, mounted a number of televised Public Service Announcements, held workshops, and participated in panel discussions with a number of other stakeholders.
8. The Bahamas Sexual Offences and Domestic Violence Act (1991). The Belize Protection Against Harassment Act (1996). The Guyana Prevention of Discrimination Act (1997). The St. Lucia Equality of Opportunity And Treatment In Employment and Occupation (2000). The Trinidad and Tobago Equal Opportunity Act (2000).
9. The Trinidad and Tobago Equal Opportunity Act No. 69 of 2000 (Part III: Sub-sect. 8–11).
10. The procedural format for conciliation varies by country: The CARICOM model calls the establishment of Tribunal "in the opinion of the authorised officer, the matter cannot be settled by conciliation" (6:1(a)). Similarly, the Trinidad and Tobago Equal Opportunity Act requires that a formal complaint be filed within six months and through the establishment of a Commission work to resolve the matter by conciliation. Even within Belize's comprehensive approach, there is a move to "carry out investigations in relation to the act and endeavour by conciliation to effect a settlement in the matter to which the act relates" (Part III: Sub-sect. 11).
11. Guyana's legislation is a notable exception to the conciliation process. It not only bypasses this process completely, but once the complainant is able to present a prima facie case of discrimination, burden of proof shifts to the respondent to disprove allegations (Part IX: Sub-sect. 23).
12. It is perplexing that the authors use a sailor's "declaration" that his "penis worked like a drill motor" as an example of an earnest (benevolent) form of sexual harassment.
13. C.O. Williams is a well-known and extremely lucrative construction company in Barbados.
14. I have recrafted the respondent's actual words to avoid any repercussions in the event I unwittingly reproduced a turn of phrase peculiar to her.

NOTES TO CHAPTER 6

1. http://www.nytimes.com/2006/10/14/nyregion/14attack.html?r=1&scp=1&sq=hate%20crime%20michael%20sandy&st=cse (accessed November 6, 2009).

2. Jurors later reported that they did not necessarily believe this to be a hate crime. They felt that the assailants were not necessarily motivated by hate or animosity toward homosexuals. Prosecutors surmised that the assailants entrapped their victim in a gay chat room under the presumption that such an individual would be less likely to resist: (http://query.nytimes.com/gst/fullpage.html?res=9C07EEDC173CF931A25753C1A9619C8B63&sec=&spon=&pagewanted=1) (accessed November 6, 2009).
3. Where I use the term "queer," which is my preferred term at this moment, I am informed by David Halperin's broad designation of queer as whatever is at "odds with the normal, the legitimate, the dominant. There is nothing in particular to which it necessarily refers. It is an identity without an essence. 'Queer' then, demarcates not a positivity but a positionality vis-à-vis the normative" (1995, 62). It is my preferred term in this project because I think it is politically productive to destabilize the categories of lesbian, bisexual, gay, and transgender (LGBT) even as important claims are presently being made in their name.
4. There is an underlying philosophical tension if not contradiction in this project that at this point I see to be irresolvable based on my desires. Even as I argue for expanding, through a range of contestations and genealogical readings, what the category "human" can mean, I do so primarily because of what liberalism has given to those to whom the category "human" has applied (citizenship, safety, security). Here lies the problem: while centering the vantage point of non-normative positionalities challenges narrow renderings of the categories "human" and "person," what is left intact are the philosophical underpinnings (e.g., individualism, rhetoric of inclusivity) that were initially exclusionary. In other words, we still stand under the sign of human and, at best, we can only move tactically within power.
5. Cornwall et al. also incorporate the emphasis of marriage normativity as part of the normalizing discourse of development (2008, 11).
6. See, for example, Greg Johnson's "The Situated Self and Utopian Thinking" (2002) for its discussion of Seyla Benhabib's arguments for the moral and political imperative of utopian thinking as well as a discussion of how the body enables such utopian envisioning. Butler's discussion of the materialization of the body through regulation is also instructive (1993).
7. The framing of women's rights as human rights is most widely attributed to Hillary Clinton who made this declaration in her speech at the Fourth World Conference on Women in 1995. More formally, the integration of a human rights-based approach within the UN framework began under the auspices of Secretary General Kofi Annan at the start of his term in 1997. Despite the significance of this human rights mainstreaming approach, there is concern that implementation and conceptual rigor have not kept pace with rhetorical fervor. See Mac Darrow and Amparo Tomas's discussion in "Power, Capture, and Conflict: A Call for Human Rights Accountability in Development Cooperation (2005, 485).
8. http://www.undg.org/archive_docs/6959-The_Human_Rights_Based_Approach_to_Development_Cooperation_Towards_a_Common_Understanding_among_UN.pdf (accessed June 15, 2010).
9. NGOs may be accorded consultative status by the UN ECOSOC. This recognition acknowledges that the NGO is representative, has a degree of expertise which the ECOCOC may draw on, and allows NGOs with such status to attend international meetings (http://esango.un.org/paperless/Web?page=static&content=intro) (accessed June 13, 2010). Julie Mertus offers an important overview of ILGA's application to ECOSOC, which was initially rejected because of member organizations within ILGA's

grouping that endorsed "inter-generational sex." This rejection persisted after ILGA expelled the group from its ranks, and their bid was again rejected in 2006 when they re-applied jointly with a Danish lesbian/gay organization (Mertus 2007).
10. The "Resolution on Human Rights and Sexual Orientation" introduced by Brazil's delegation to the UN in 2003 affirmed the importance of applying the principles of human rights to sexual minorities (E/CN.4/2003/L.92 http://daccess-dds-ny.un.org/doc/UNDOC/LTD/G03/138/18/PDF/G0313818.pdf?OpenElement (accessed June 14, 2010). The 2008 "Statement on Human Rights, Sexual Orientation and Gender Identity" also affirmed human rights principles, voiced concern for violations of these rights, called for cessation of human rights violations toward sexual minorities, and called on the Human Rights Commission to address these violations (A63/635 http://www.sxpolitics.org/wp-content/uploads/2009/03/un-document-on-sexual-orientation.pdf (accessed June 15, 2010).
11. Merry identifies the use of local recognizable symbols and images in vernacularization as "indigenization." She notes, however, that because the philosophical premises of a human rights framework take precedence in vernacularization, it is sometimes not possible to indigenize completely (2006, 48). This is a troubling observation because it would suggest that the local must then defer to global renderings of "rights." This smacks of the hegemony that makes many in the global South wary. Here, it may be useful to read Shaheen Sardar Ali's "Women's Rights, CEDAW, and International Human Rights Debate toward Empowerment?" (2002).
12. Please see Yasmin Tambiah's "Creating (Im)moral Citizens: Gender Sexuality and Lawmaking in Trinidad and Tobago, 1986" (2009). Drawing on research data that Alexander did not have access to at the time of her research, Tambiah reinterrogates and builds on some of Alexander's earlier argument. Of note, Tambiah argues that lesbian sexuality was not necessarily singled out by the state, but rather, a casualty of "gender neutral" language. I see Tambiah's work as an important companion piece to Alexander's earlier discussion.
13. In her later work, she refers to this kind of temporal interplay, with its symbolic effects, as palimpsestic, "a parchment that has been inscribed two or three times, the previous text having been imperfectly erased and remaining therefore still partly visible" (2005, 190).
14. See Kamala Kempadoo's discussion of a sexuality praxis as a means of interrogating policy (2009).
15. Mati work, from which "matism" is derived, refers to a sexual and economic arrangement wherein Afro-Surinamese women "engage in sexual relationships with people of the opposite gender and people of the same gender, either consecutively or simultaneously" (Wekker 1997, 331). Drawing on a larger religious cosmology, it does not link the activity of matism to an identity (Wekker 2006, 194).
16. See, as part of a Caribbean queer etymology, Audre Lorde's discussion of the term "zami" (1982) and Wesley E. Crichlow's discussion of the importance of reclaiming the term "bullerman" (2004) as examples of what I am referring to here as the work of symbolic capture.
17. One criterion for proving credibility of one's petition for asylum to the United States is that one has to convince an immigration judge that the conditions of life in the home territory are severe: http://www.immigrationequality.org/blog/?p=1469 (accessed June 11, 2010).
18. The global gaze on the region due to the repeated statistic that the region has the second highest infection rate for HIV/AIDS (after Africa) not only invokes old racialized bogeys but also opens scrutiny to the state's management of MSM as part of the global management of public health. This scrutiny

Notes   219

does not necessarily emanate out of a progressive politics but essentially aims to regulate the perceived contaminated queer subject. See http://www.nytimes.com/2005/11/30/opinion/30wed3.html (accessed June 11, 2010).
19. Cecil Gutzmore (2004), for example, identifies five components that construct a particular discourse about homosexuality in Jamaica, much of which would have preceded Banton. These include religious fundamentalism, homosexuality as "unnatural," narratives of homosexuality as antithetical to African culture, conflation of homosexuality and pedophilia, and the illegality of homosexuality.
20. This designation of superlative homophobia now circulates in a number of mainstream locations. See, for example, "The Most Homophobic Place on Earth?" in *Time Magazine* (April 12, 2006) at http://www.time.com/time/world/article/0,8599,1182991,00.html (accessed June 11, 2010). See also the episode of the *Oprah* show, "Gay Around the World"; the episode examined instances of global homophobia and featured New York-based Jamaican poet Stacy Ann Chin. An entire conference held at Nova University in Florida in March 2010 was dedicated to the Caribbean's homophobia: http://nsunews.nova.edu/homophobia-caribbean-explored-symposium-nsu/ (accessed April 12, 2010).
21. See my colleague Katie King's discussion in "'There Are No Lesbians Here': Lesbianisms, Feminisms, and Global Gay Formations" (2002).
22. Jamaica Forum for Lesbians, All Sexualities and Gays (J-FLAG), United Gay and Lesbian Association of Barbados (UGLABB), Trinidad and Tobago Coalition Advocating for Inclusion of Sexual Orientation (CAISO), Guyana Rainbow Foundation (GUYBOW), SASOD, and Caribbean Forum for Liberation and Acceptance of Genders and Sexualities (CARIFLAGS).
23. In 2001, the Guyanese Parliament approved a constitutional amendment that would make it illegal to discriminate based on sexual orientation. The Amendment was subsequently sent back to Parliament by the President after he refused to assent, as reported in Patrick Denny's "Sexual Orientation Bill Going Back to Parliament" (*Stabroek News* January 26, 2001).
24. I am always amused when I learn from SASOD about changes to legislation that affect queer communities in the United States, where I reside, before learning of it in the news "here." The most recent example of SASOD's disruption of here/there was the announcement of the U.S. State Department's changes to the passport guidelines that require individuals who are applying for a change of gender status on their passport to submit certification from a medical provider that they have received "appropriate clinical treatment" without needing to show evidence of gender reassignment surgery. See http://www.state.gov/r/pa/prs/ps/2010/06/142922.htm (accessed May 14, 2010).
25. http://sasod.blogspot.com/2007/03/sasod-challenges-ministry-of-health.html (accessed June 5, 2010).
26. http://sasod.blogspot.com/2007/03/sasod-challenges-ministry-of-health.html (accessed June 12, 2010).
27. Additionally, the film festival is free, centrally located, accessible by affordable public transportation, and while such possibilities are not necessarily realizable or advised in the variegated hostilities of the region, it is critical to the transformation of public space. However, it is important to note that a queer geography should not be limited to only those spaces of public place making.
28. http://www.mulabi.org/epu/8th%20round/Guyana%20Report%20-%20SRI%20&%20SASOD.pdf (accessed June 13, 2010). It is also important to note that at the point that this project went to press, Guyana successfully secured gender-neutral language regarding rape with the passage of the Sexual Offences Act 2010 (Section 3). See: http://www.parliament.gov.gy/documents/sexact2010.doc (accessed June 20, 2010).

29. I am using variegated hostilities as a reminder that the experience and expression of homophobia differs from one Caribbean territory to the other.
30. http://www.stabroeknews.com/2009/news/courts/02/10/he-wore-blue-velvetseven-fined-for-cross-dressing/ (accessed June 13, 2010).
31. http://www.trinidadexpress.com/index.pl/article_news?id=161601654 (accessed April 13, 2010).
32. http://sasod.blogspot.com/ (accessed April 13, 2010).
33. See, for example, Mary L. Gray's "Negotiating Identities/Queering Desires: Coming Out Online and the Remediation of the Coming Out Story," in *Journal of Computer Mediated Communication* 14 (2009): 1162–1189.
34. http://www.stabroeknews.com/2010/features/03/01/this-case-is-about-and-for-all-of-us (accessed June 17, 2010).
35. See http://www.stabroeknews.com/2010/letters/02/27/nothing%e2%80%98historic%e2%80%99-about-cross-dressing-motion/ (accessed June 17, 2010).

Bibliography

Aaltio, Iris, and Albert J. Mills. *Gender, Identity and the Culture of Organizations.* New York: Routledge, 2002.
Ahmed, Aziza, with DAWN Caribbean Steering Committee. *Maternal Mortality, Abortion, and Health Sector Reform in Four Caribbean Countries: Barbados, Jamaica, Suriname and Trinidad and Tobago.* Caribbean: DAWN Sexual and Reproductive Rights Health Program, 2005.
AIDS-Free World. *Electing to Rape: Sexual Terror in Mugabe's Zimbabwe* (December 2009). http://www.aids-freeworld.org/images/stories/zimbabwe/zim%20grid%20screenversionfinal.pdf (accessed May 17, 2010).
Albrow, Martin. *Do Organizations Have Feelings?* London and New York: Routledge, 1997.
Alcoff, Linda. "Cultural Feminism versus Post-Structuralism: The Identity Crisis in Feminist Theory." *Signs* 13, no. 3 (1988): 405–436.
Alexander, Jacqui M. "Redrafting Morality: The Postcolonial State and the Sexual Offenses Bill of Trinidad and Tobago." In *Third World Women and the Politics of Feminism*, eds. Chandra Mohanty, Ann Russo, and Lourdes Torres, 133–152. Indiana: Indiana University Press, 1991.
———. "Not Just (Any)Body Can be a Citizen: The Politics of Law, Sexuality and Postcoloniality in Trinidad and Tobago and the Bahamas." *Feminist Review* 48 (1994): 5–23.
———. Erotic Autonomy as a Politics of Decolonization: An Anatomy of Feminist and State Practice in the Bahamas Tourist Economy. In *Feminist Genealogies, Colonial Legacies, Democratic Futures*, eds. M. Jacqui Alexander and Chandra Mohanty, 63–100. New York: Routledge, 1997.
———. *Pedagogies of Crossing: Meditations on Feminism, Sexual Politics, Memory, and the Sacred.* Durham: Duke University Press, 2005.
Ali, Shaheen Sardar. "Women's Rights, CEDAW, and International Human Rights Debate toward Empowerment?" In *Rethinking Empowerment, Gender and Development in a Global/Local World*, ed. Jane Parpart et al., 61–78. New York: Routledge, 2002.
Alston, Phillip. "Ships Passing in the Night: The Current State of the Human Rights and Development Debate Seen through the Lens of the Millennium Development Goals." *Human Rights Quarterly* 27 (2005): 755–829.
Alvarez, Sonia. "Translating the Global: Effects of Transnational Organizing." *Meridians: Feminism, Race, Transnationalism* 1, no. 1 (2000): 29–67.
Altman, Dennis. *Global Sex.* Chicago: University of Chicago Press, 2001.
Anan, Kofi. "Message to the Fifth Asian and Pacific Population Conference (Ministerial Segment). Delivered by Mr. Kim Hak-Su, Executive Secretary, UNESCAP. Bangkok, 16 December, 2002.

Anderson, M. et al. "Liminal Identities: Caribbean Men Who Have Sex with Men in London, U.K." *Culture, Health and Sexuality* 11, no. 3 (2009): 315–330.

Antoine, Rose-Marie. "Charting a Legal Response to HIV and AIDS and Work from the Perspective of Vulnerability" In *Sexuality, Social Exclusion and Human Rights,* eds. Christine Barrow, Marjan de Bruin, and Robert Carr, 38–71. Kingston, Jamaica: Ian Randle Press, 2009.

Antrobus, Peggy. "Women in Development Programmes: The Caribbean Experience (1975–1985)." In *Gender in Caribbean Development*, eds. Patricia Mohammed and Cathy Shepherd, 36–54. The University of the West Indies: UWI Women and Development Studies Project, 1988.

———. "Critiquing the MDGs from a Caribbean Perspective." *Gender and Development* 13, no. 1 (2005): 94–104.

Appadurai, Arjun. *The Social Life of Things: Commodities in Cultural Perspective*. Cambridge: Cambridge University Press, 1988.

———. *Modernity at Large: Cultural Dimensions of Globalization*. Minnesota: University of Minnesota Press, 1996.

Armas, H. "Whose Sexualities Count?: Poverty, Participation, and Sexual Rights." *IDS Working Paper 294*. Brighton: Institute of Development Studies, 2007.

Armstrong, Pat. "The Feminization of the Labour Force: Harmonizing Down in a Global Economy." In *Rethinking Restructuring: Gender and Change in Canada*, ed. I. Bakker, 29–54. Toronto: Toronto University Press, 1996.

Bacchi, Carol. "Gender Mainstreaming: A New Vision, More of the Same, or Backlash?" *Dialogue: Academy of the Social Sciences* 20, no. 2 (2001): 16–20.

Bacchi, Carol, and Joan Eveline. "What Are We Mainstreaming When We Mainstream Gender?" *International Feminist Journal of Politics* 7, no. 4 (2005): 496–512.

Baden, Sally, and Anne Marie Goetz. "Who Needs [Sex] When You Can Have [Gender]? Conflicting Discourses on Gender at Beijing." In *Feminist Visions of Development, Gender, Analysis and Policy*, eds. Cecile Jackson and Ruth Pearson, 19–37. New York: Routledge, 1998.

Bailey, Barbara. "Sexist Patterns of Formal and Non-Formal Education Programmes: The Case of Jamaica." In *Gender: A Caribbean Multi-Disciplinary Perspective*, ed. Elsa Leo-Rhynie, et al., 144–158. Jamaica: Ian Randle Publishers, 1997.

———. "Feminism and Educational Research and Understandings: The Stae of the Art in the Caribbean." In *Caribbean Portraits: Essays on Gender Ideologies and Identities*, ed Christine Barrow, 208–224. Jamaica: Ian Randle Publishers, 1998.

———. "Gendered Realities: Fact or Fiction? The Realities in a Secondary Level Co-educational Classroom." In *Gendered Realities: Essays in Caribbean Feminist Thought*, ed Patricia Mohammed, 104–182. Jamaica: UWI Press and CGDS, Mona, 2002.

Bakhtin, M.M. *The Dialogic Imagination: Four Essays*. Translated by Caryl Emerson and Michael Holquist. Texas: University of Texas Press, 1982.

Barbados Police Act 1908–02 (ammended 1955).

Barriteau, Eudine. "Postmodernist Feminist Theorizing and Development Policy and Practice in the Anglophone Caribbean: The Barbados Case." In *Feminism/Postmodernism/Development*, eds. Jane Parpart and Marianne Marchand, 142–158. London: Routledge, 1995.

———. "Feminist Theory and Development: Implications for Policy, Research and Action." In *Theoretical Perspectives on Gender and Development*, eds. Jane L. Parpart, M. Patricia Connelly, and V. Eudine Barriteau, 161–177. Ottawa, Canada: IDRC, 2000.

———. "Women Entrepreneurs and Economic Marginality: Rethinking Caribbean Women's Economic Relations." In *Gendered Realities: Essays in Caribbean Feminist Thought*, ed. Patricia Mohammed, 221–248. Mona, Jamaica: University of the West Indies Press, 2002.

———. "Requiem for the Male Marginalization Thesis in the Caribbean: Death of a Non-Theory." In *Confronting Power, Theorizing Gender: Interdisciplinary Perspectives in the Caribbean*, ed. Eudine Barriteau, 324–355. Mona, Jamaica: University of the West Indies Press, 2003.

———. "Constructing Feminist Knowledge in the Commonwealth Caribbean in the Era of Globalization." In *Gender in the 21st Century: Caribbean Perspectives, Visions and Possibilities*, eds. Barbara Bailey and Elsa Leo-Rhynie, 437–465. Kingston: Ian Randle Publishers, 2004.

Barrow, Christine. *Family in the Caribbean: Themes and Perspectives*. Kingston: Ian Randle Publishers and Oxford: James Currey Publishers, 1996.

Basu, Amrita. "Globalization of the Local/Localization of the Global." *Meridians: feminism, race, transnationalism* 1, no. 1 (2000): 68–84.

Baugh, Edward. "The Leader Speaks. (People Power)" *A Tale from the Rainforest*. Toronto, Canada: Sandberry Press, 1988.

Beall, Jo. "'Trickle-down or Rising Tide?' Lessons on Mainstreaming Gender Policy from Columbia and South Africa." *Social Policy Administration* 32, no. 5 (December 1998): 513–534.

Beckles, Hilary. *Natural Rebels: A Social History of Enslaved Black Women in Barbados*. New Jersey: Rutgers University Press, 1989.

———. *Centering Woman: Gender Discourses in Caribbean Slave Society*. Kingston: Ian Randle Publishers, 1999.

Berkovitch, Nitza. *From Motherhood to Citizenship: Women's Rights and International Organizations*. Baltimore, MD: The John Hopkins University Press, 1999.

Berlant, Lauren. *The Queen of America Goes to Washington: Essays on Sex and Citizenship*. Durham:Duke University Press, 1997.

Bhabha, Homi. *The Location of Culture*. New York: Routledge, 1994.

Bhana, Deevia. "Children's Sexual Rights in an Era of HIV/AIDS." In *Development with a Body: Sexuality, Human Rights and Development*, ed. Andrea Cornwwall et al., 77–85. London: Zed Books, 2008.

Binnie, Jon. *The Globalization of Sexuality*. London: SAGE. 2004.

Blake, Judith. *Family Structures in Jamaica: The Social Context of Reproduction*. New York: The Free Press of Glencoe, 1961.

Blomstrom, Magnus, and Bjorn Hettne, *Development Theory in Transition. The Dependency Debate and Beyond: Third World Responses*. London: Zed Books, 1984.

Bolles, Lynn. *We Paid Our Dues: Trade Union Leaders of the Caribbean*. Washington D.C.: Howard University Press, 1996.

Bordo, Susan. "The Body and the Reproduction of Femininity." In *Writing on the Body: Female Embodiment and Feminist Theory*, eds. Katie Conboy, Nadia Medina, and Sarah Stanbury, 90–110. New York: Columbia University Press, 1997.

Boserup, Ester. *The Conditions of Agricultural Growth: The Economics of Agrarian Change under Population Pressure*. London: Unwin and Allen, 1965.

———. "Obstacles to Advancement of Women During Development." In *Investment in Women's Human Capital*, ed. T. Paul Schultz, 51–60. Chicago: University of Chicago Press, 1995.

Bourne, Compton. "A Caribbean Community for All." *Fifth Lecture in the Distinguished Lecture Series Celebrating the Thirtieth Anniversary of the Caribbean Community*. Grenada: Caribbean Development Bank, 2003.

Braidotti, Rosi. *Nomadic Subjects*. New York: Columbia University Press, 1994.

Branca, Patricia. *Silent Sisterhood: Middle Class Women in the Victorian Home.* London: Croom Helm, 1975.
Brereton, Bridget. *Race Relations in Colonial Trinidad: 1870–1900.* Cambridge: Cambridge University Press, 1979.
———."General Problems and Issues in Studying the History of Women." In *Gender in Caribbean Development*, eds. Patricia Mohammed and Cathy Shepherd, 125–143. The University of the West Indies: Women and Development Studies Project, 1988.
———. "The Development of an Identity: The Black Middle Class of Trinidad in the Late Nineteenth Century." In *Caribbean Freedom. Economy and Society: From Emancipation to the Present, A Student Reader*, eds. Hilary Beckles and Verene Shepherd, 281–282. Kingston, Jamaica: Ian Randle Publishers and London: James Currey Publisher, 1993.
———. *Gendered Testimony: Autobiographies, Diaries, and Letters by Women as Sources for Caribbean History.* The 1994 Elsa Goveia Memorial Lecture. The University of the West Indies, Mona: Department of History, 1994.
———. "Family Strategies, Gender, and the Shift to Wage Labour in the British Caribbean." *Journal of Social Sciences* 4, no. 2 (1997.): 32–55.
Britzman, Deborah P. "'The Question of Belief': Writing Poststructural Ethnography." In *Working the Ruins: Feminist Poststructural Theory and Methods in Education*, eds. E. St. Pierre and W. Pillow, 27–40. New York: Routledge, 1999.
Brown, Wendy. "Finding the Man in the State." *Feminist Studies* 18, no. 1 (Spring 1992): 7–34.
Brown-Campbell, Gladys. *Patriarchy in the Jamaica Constabulary Force: Its Impact on Gender Equality.* Kingston, Jamaica: Canoe Press, 1998.
Bush, Barbara. *Slave Women in Caribbean Society 1650–1838.* Kingston, Jamaica: Ian Randle Publishers and Bloomington: Indiana University Press Bloomington, 1990.
Busia, Abena. "Silencing Sycorax: On Colonial Discourse and the Unvoiced Female." *Cultural Critique* 14 (1990): 81–104.
Butler, Judith. "Performative Acts and Gender Constitution: An Essay in Phenomenology and Feminist Theory," *Theatre Journal* 40, no. 4 (1988), pp. 519–531.
———. *Gender Trouble: Feminism and the Subversion of Identity.* New York: Routledge, 1990.
———. *Bodies That Matter: On the Discursive Limits of Sex.* New York: Routledge, 1993.
———. *Undoing Gender.* New York: Routledge, 2004.
Caribbean Development Bank. *Achieving the Millennium Development Goals in Borrowing Member Countries: The Role of the Special Development Fund and the Caribbean Development Bank.* Barbados: Caribbean Development Bank, 2005.
Carney, Gemma. "Communicating or Just Talking? Gender Mainstreaming and the Communication of Global Feminism" *Women and Language*, XXVI, no. 1 (2003): 52–60.
Charles, A.E. "Abortion in Trinidad and Tobago." Unpublished manuscript, April 30, 2000.
Charlesworth, Hilary. "Not Waving but Drowning: Gender Mainstreaming and Human Rights in the United Nations." *Harvard Human Rights Journal* 18 (2005): 1–18.
Chevannes, Barry. *Learning to Be a Man. Culture, Socialization, and Gender Identity in Five Caribbean Communities.* Kingston, Jamaica: University Press of the West Indies, 2001.
Chin, Timothy. "Jamaican Popular Culture, Caribbean Literature and the Representation of Gay and Lesbian Sexuality in the Discourses of Race and Nation." *Small Axe* 5(1999): 14–33.

———. "Bullers and Battyman: Contesting Homophobia in Black Popular Culture and Contemporary Caribbean Literature." *Callaloo* 20, no. 1 (1997): 127–41.

Chowdhury, Elora. "Feminist Negotiations: Contesting Narratives of the Campaign against Acid Violence in Bangladesh" *Meridians: feminisms, race, transnationalism* 6, no. 1 (2005): 163–92.

Coalition against All Forms of Sexual Harassment. *Charting Gender Equality in the Workplace: With a Special Emphasis on Sexual Harassment. Final Report.* CASH: Barbados, 2005.

Connell, R.W. "Change among the Gatekeepers: Men, Masculinities and Gender Equality in the Global Arena." *Signs: Journal of Women, Culture and Society* 30, no. 3 (Spring 2005): 116–133.

Cook, Rebecca J., and Bernard M. Dickens. "Human Rights Dynamics of Abortion Law Reform." *Human Rights Quarterly* 25 (2003): 1–59.

Cornwall, Andrea et al. *Development with a Body: Sexuality, Human Rights and Development.* London: Zed Books, 2008.

Correa, Sonia, and Muntarbhorn, V. 2007. The Yogyakarta Principles on the Application of International Human Rights Law in Relation to Sexual Orientation and Gender Identity. http://www.yogyakartaprinciples.org/principles_en.htm (accessed May 22, 2010).

Craton, Michael. *Searching for the Invisible Man: Slaves and Plantation Life in Jamaica.* Cambridge: Harvard University Press, 1978.

Crichlow, Wesley E. "History, (Re)Memory, Testimony and Biomythography: Charting a Buller Man's Trinidadian Past." In *Interrogating Caribbean Masculinities: Theoretical and Empirical Analyses*, ed. Rhoda E. Reddock, 185–222. Kingston, Jamaica: The University of the West Indies Press, 2004.

Darrow, Mac, and Tomas Amparo. "Power, Capture, and Conflict: A Call for Human Rights Accountability in Development Cooperation. *Human Rights Quarterly* 27, no. 2 (May 2005): 471–538.

Desai, Manisha. "Transnationalism: The Face of Feminist Politics Post-Beijing." *International Social Science Journal* 17, no 184 (2005): 319–30.

Douglas, Mary. *Natural Symbols: Explorations in Cosmology.* 2nd edition. New York: Routledge, 2003.

Dunlop, Joan et al. "Women Redrawing the Map: The World after the Beijing and Cairo Conferences." *SAIS Review* 6, no. 1 (Winter–Spring 1996): 153–165.

Dutt, Malika. "Reclaiming a Human Rights Culture: Feminism of Difference and Alliance." In *Talking Visions: Multicultural Feminism in a Transnational Age.* ed. Ella Shoat, 225–246. New York: New Museum of Contemporary Art, 1998.

Eisenstein, Hester. *Inside Agitators: Australian Femocrats and the State.* Philadelphia: Temple University Press, 1996.

El-Bushra, Judy. "Rethinking Gender and Development Practice for the Twenty-first Century." *Gender and Development* 8, no. 1 (March 2000): 55–62.

Ellis, Pat. *Women in the Caribbean.* London: Zed Books, 1986.

Elson, Diane. "Talking to the Boys: Gender and Economic Growth Models." In *Feminist Visions of Development: Gender, Analysis and Policy*, eds. Cecile Jackson and Ruth Pearson, 155–170. New York: Routledge, 1998.

Enloe, Cynthia. *Bananas, Beaches and Bases: Making Feminist Sense of International Politics.* Los Angeles: University of California Press, 2000.

Escobar, Arturo. "Discourse and Power in Development: Michel Foucault and the Relevance of his Work to the Third World." *Alternatives* 10, no. 3 (1984): 377–400.

———. *Encountering Development: The Making and Unmaking of the Third World.* Princeton, New Jersey: Princeton University Press, 1995.

Family Care International. *Millennium Development Goals Sexual and Reproductive Health: Briefing Cards*, 2005. http://www.familycareintl.org/userfiles/file/pdfs/MDG-cards-AN.pdf (accessed June 1, 2010).

Fanon, Frantz. *Black Skin, White Mask*. New York: Grove Press, 1967.

FAO Corporate Document Repository: *The Barbados Food Consumption and Anthropometric Surveys 2000. Prepared in collaboration with the FIVIMS Secretariat by a team from the National Nutrition Centre, Ministry of Health, Government of Barbados*. Food and Agriculture Organization of the United Nations: Rome, 2005.http://www.fao.org/documents/show_cdr.asp?url_file=/docrep/008/y5883e/y5883e09.htm (accessed May 25, 2007).

Fenstermaker, Sarah, and Candace West. *Doing Gender, Doing Difference: Inequality, Power, and Institutional Change*. New York: Routledge, 2002.

Figueroa, Mark. "Challenging Gender Privileging: A Caribbean Experience." Paper presented at the Mona Academic Conference *Gender in the 21st Century: Perspectives, Visions and Possibilities*, The University of the West Indies (Mona), Centre for Gender and Development Studies, Regional /Coordinating Unit, Mona Campus, August 29–31 2003.

Fiske, Susan T., and Glick, P. "Ambivalence and Stereotypes Cause Sexual Harassment: A Theory with Implications for Organizational Change." *Journal of Social Issues*, 51, no. 1 (1995): 97–115.

———. The Ambivalent Sexism Inventory: Differentiating Hostile and Benevolent Sexism. *Journal of Personality and Social Psychology*, 70 (1996): 491–512.

Flax, Jane. *Disputed Subjects: Essays on Psychoanalysis, Politics, and Philosophy*. New York: Routledge, 1993.

Foucault, Michel. *The Archaeology of Knowledge and the Discourse on Language*. Translated by A.M. Sheridan Smith. New York: Pantheon Books, 1972.

———. "Human Nature: Justice versus Power." In *Reflexive Water: The Basic Concerns of Mankind*, ed. Fons Elder, 184–185. London: Souvenir Press, 1974.

———. *Power/Knowledge: Selected Interviews and Other Writings, 1972–1977*. ed. by Colin Gordon. New York: Pantheon, 1980.

———. *The History of Sexuality. Volume 1: An Introduction*. Translated by Robert Hurley. New York: Vintage Books, 1990.

———. *Discipline and Punish: The Birth of the Prison*. Translated by Alan Sheridan. New York: Vintage Books, 1995.

———. "On the Ways of Writing History? In *Michel Foucault: Power. Essential Works of Foucault 1954–1984. Vol. 2*, ed. James Faubion, 261–95. New York: The New Press, 1999.

———. *Michel Foucault: Power. Essential Works of Foucault 1954–1984 Series Vol. 3*, ed. James D. Faubion, 223–38. New York: The New Press, 2000.

Franke, Katherine. "What's Wrong with Sexual Harassment?" *Stanford Law Review* 49, no. 4 (April 1997): 691–772.

Fraser, Arabella. "Approaches to Reducing Maternal Mortality: Oxfam and the MDGs." *Gender and Development* 13, no. 1 (2005): 36–43.

Freeman, Carla. *High Tech and High Heels in the Global Economy: Women, Work, and Pink Collar Identities in the Caribbean*. Durham: Duke University Press, 2000.

Friedman, Elizabeth. "The Effects of 'Trans-nationalism Reversed' in Venezuela: Assessing the Impact of UN Global Conferences on the Women's Movement." *International Feminist Journal of Politics* 1, no. 3 (Autumn 1995): 357–81.

Ghany, Hamid. "Parliament in Trinidad and Tobago 1962–1987. The Whitehall Model at Work." In *Trinidad and Tobago: The Independence Experience 1962–1987*, ed. Selwyn Ryan, UWI, St. Augustine, Trinidad: I.S.E.R, 1988.

Glick, Peter. "Trait-Based and Sex-Based Discrimination in Occupational Prestige, Occupational Salary, and Hiring." *Sex Roles* 25, no. 5–6 (1991): 351–78.
Goetz, Anne Marie. *Getting Institutions Right for Women in Development*. London: Zed Books, 1997.
———."National Women's Machinery: State-Based Institutions to Advocated for Gender Equality. In *Mainstreaming Gender, Democratizing the State: Institutional Mechanisms of the Advancement of Women*, ed. Shirin M. Rai, 69–95. New Brunswick: Transaction Publishers, 2007.
Gosine, Andil. *Sex for Pleasure, Rights to Participation, and Alternatives to AIDS: Placing Sexual Minorities and/or Dissidents in Development IDS Working Paper 228*. Brighton, England: Institute of Development Studies, 2005.
———. "'Race,' Culture, Power, Sex, Desire, Love: Writing in 'Men Who Have Sex with Men.'" *IDS Bulletin* 37, no. 5 (2007): 27–33.
———. "Speaking Sexuality: The Heteronationalism of MSM." In *Sexuality, Social Exclusion, and Human Rights*, eds. Christine Barrow, Marjan de Bruin, and Robert Carr, 95–115. Kingston, Jamaica: Ian Randle Press, 2009.
Gouws, Amanda. "The Rise of the Femocrat?" *Agenda* 30 (1996): 31–43.
Goveia, Elsa. *Amelioration to Emancipation*. The University of the West-Indies, St. Augustine: History Department, 1977.
Government of Barbados. *The House of Assembly Debates (Official Report) First Session 1981–1986*. Government Printery. Bridgtown: Barbados, 1983.
Government of Trinidad and Tobago. *National Report on the Status of Women in Trinidad and Tobago. Prepared for the Fourth World Conference on Women Beijing, China, September 4–15, 1995*. Gender Affairs Division of Trinidad and Tobago: Trinidad and Tobago, 1995.
———. *Population and Vital Statistics Report*. Central Statistical Office: Trinidad and Tobago, 1999.
———. *Initial, Second, and Third Periodic Report of the Republic of Trinidad and Tobago, International Convention on the Elimination of Discrimination against Women*. Ministry of the Attorney General and Legal Affairs: Trinidad and Tobago, 2000.
———. "Finance Committee Report Trinidad and Tobago Hansard, House of Representatives, September 22, 2000.
———. *Budget of Trinidad and Tobago: Vision 2020 Ensuring Our Future Prosperity Addressing Basic Needs 2006. Republic of Trinidad and Tobago*. http://www.finance.gov.tt/documents/publications/pub7.pdf (accessed August, 23, 2008)
Gutek, Barbara et al. "Predicting Social-Sexual Behavior at Work: A Contact Hypothesis." *Academy of Management Journal* 33, no. 3 (1990): 560–577.
———, and Bruce Morasch. "Sex-Ratio and Sex-Spill over and Sexual Harassment at Work." *Journal of Social Issues* 38 (Winter 1982): 55–74.
Gutzmore, Cecil. "Casting the First Stone! Policing of Homo/Sexuality in Jamaican Popular Culture." *Interventions* 6, no. 1 (2004): 118–134.
Halberstam, Judith. *Female Masculinity*. Durham: Duke University Press, 1998.
Halperin, David. *Saint Foucault: Towards a Gay Hagiography*. New York: Oxford University Press, 1995.
Harcourt, Wendy. *Body Politics in Development: Critical Debates in Gender and Development*. London: Zed Books, 2009.
Harcourt, Wendy, and Arturo Escobar, eds. *Women and the Politics of Place*. Bloomfield, Connecticut: Kumarian Press, 2005.
Harris, Sonja T. "Review of Institutional Mechanisms for the Advancement of Women and for Achieving Gender Equality 1995–2000." In *Gender Equality in the Caribbean: Reality or Illusion*, eds. Gemma Tang Nain and Barbara Bailey, 178–200. Kingston: Ian Randle Publishers, 2003.

Hayes, Ceri. "Out of the Margins: The MDGs through a CEDAW Lens." *Gender and Development* 13, no. 1 (March 2005) 67–78.

Heyzer, Noeleen. "Making the Links: Women's Rights and Empowerment Are Key to Achieving the Millennium Development Goals." *Gender and Development* 13, no. 1 (March 2005): 9–12.

Higman, Barry. *Slave Populations of the British Caribbean 1807–1834*. Kingston, Jamaica: UWI Press, 1995.

Hoffman, Elizabeth A. "Selective Sexual Harassment: Differential Treatment of Similar Groups of Women Workers." *Law and Human Behavior* 28, no. 1 (2004): 29–45.

Holland, Janet et al. "Power and Desire: The Embodiment of Female Sexuality." *Feminist Review*, 46 (Spring 1994): 21–38.

IDB (Inter-American Development Bank). *Non-Traditional Skills Training Program*. Project No. TC-94-04-37-1.

Jackson, Cecile, and Ruth Pearson. *Feminist Visions of Development: Gender, Analysis, and Policy*. London: Routledge, 1998.

Jahan, Rounaq. *The Elusive Agenda: Mainstreaming Women in Development*. London: Zed Books, 1995.

Johnson, Greg. "The Situated Self and Utopian Thinking." *Hypatia* 17, no. 3 (2003): 20–44.

Jolly, Susie. *Why the Development Industry Should Get Over Its Obsession with Bad Sex and Start to Think about Pleasure*. IDS Working Paper 283. Brighton: Institute of Development Studies, 2007.

———. "Not so Strange Bedfellows: Sexuality and International Development." *Development (Women's Rights and Development)* 49, no. 1 (January 2006): 77–80.

Jones, Kathleen B. "Towards the Revision of Politics." In *The Political Interests of Gender: Developing Theory and Research with a Feminist Face*, eds. Kathleen B. Jones and Anna G. Jonasdottir, 11–32. London: Sage Modern Politics Series, 1988.

Kabeer, Naila. *Reversed Realities: Gender Hierarchies in Development Thought*. London: Verso Press, 1994.

———. "Gender Equality and Women's Empowerment: A Critical Analysis of the Third Millennium Development Goals." *Gender and Development* 13, no. 1 (2000): 13–24.

———. Gender Mainstreaming in Poverty Eradication and the Millennium Development Goals: A Handbook for Policy-Makers and Other Stakeholders (New Gender Mainstreaming in Development Series). London: Commonwealth Secretariat, 2003.

———. "Gender Equality and Women's Empowerment: A Critical Analysis of the Third Millennium Development Goal." *Gender and Development* 13, no. 1 (March 2005) 13–24.

Kardam, Nuket, and Selma Acuner. "National Women's Machineries: Structures and Spaces." In *Mainstreaming Gender, Democratizing the State: Institutional Mechanisms for the Advancement of Women*, ed. Shirin Rai, 96–113. New Brunswick: Transaction Publishers, 2007.

Kempado. Kamala. "Caribbean Sexuality: Mapping the Field" *Caribbean Review of Gender Studies: A Journal of Caribbean Perspectives on Gender and Feminism* 3 (November 2009). http://sta.uwi.edu/crgs/november2009/journals/Kempadoo.pdf (accessed June 12, 2010).

———. "Centering Praxis in Policies and Studies of Caribbean Sexuality." In *Sexuality Social Exclusion and Human Rights: Vulnerability in the Caribbean Context of HIV*, eds. Christine Barrow, Marjan de Bruin, and Robert Carr, 179–191. Kingston, Jamaica: Ian Randle Press, 2009.

Kessler, Suzanne J. *Lessons from the Intersexed*. New Jersey: Rutgers University Press, 1998.

King, Katie. "'There Are No Lesbians Here': Lesbianisms, Feminisms, and Global Gay Formations." In *Queer Globalizations: Citizenship and the Afterlife of Colonialism*, eds. Arnaldo Cruz-Malave and Martin Manalansen IV, 33–45. New York: SUNY Press, 2002.

Lazarus-Black, Mindie. "The Rites of Domination: Tales from Domestic Violence Court." Working Paper no. 7. St. Augustine, UWI: Centre for Gender and Development Studies, 2002.

———. "The (Heterosexual) Rendering of a Modern State: Criminalizing and Implementing Domestic Violence in Trinidad." *Law and Social Inquiry* 28, no. 4 (2003): 979–1003.

Lazreg, Marnia. "Development: Feminist Theory's Cul-De-Sac." In *Feminist Post-Development Thought: Rethinking Modernity, Post-colonialism and Representation*, ed. Kriemild Saunders, 123–145. New York: Zed Books, 2002.

Leo-Rhynie, Elsa et al. *Gender: A Caribbean Multi-disciplinary Perspective*. Kingston: Ian Randle Press, 1997.

Lindsay, Keisha. "Is the Caribbean Male an Endangered Species?" In *Gendered Realities: Essays in Caribbean Feminist Thought*, ed. Patricia Mohammed, 56–82. Kingston, Jamaica: University of the West Indies Press, 2003.

Lloyd, Anthony, and Elaine Robertson. *Social Welfare in Trinidad and Tobago*. Trinidad: Antilles Research Associates, 1971.

Lloyd, Moya. *Beyond Identity Politics: Feminism, Power and Politics*. London: Sage Press, 2005.

Loomba, Ania. *Colonialism/Postcolonialism*. New York: Routledge, 1998.

Lorber, Judith. Paradoxes of Gender. New Haven, CT: Yale University Press, 1994.

Lorde, Audre. *Zami: A New Spelling of My Name—A Biomythography*. California: Crossing Press, 1983.

MacKinnon, Catharine. *Sexual Harassment of Working Women: A Case of Sex Discrimination*. New Haven, CT: Yale University Press, 1979.

———. *Toward a Feminist Theory of the State*. Cambridge, MA: Harvard University Press, 1991.

———. "Sex Equality: On Difference and Dominance." In *Theorizing Feminism: Parallel Trends in the Humanities and Social Sciences*, eds. Anne C. Herrmann and Abigail J. Stewart, 232–253. Boulder, Colorado: Westview Press, 2001.

Marchand, Marianne, and Jane Parpart. *Feminism/Postmodernism/Development*. London: Routledge, 1995.

Massiah, Joycelin. *Women as Heads of Households in the Caribbean: Family Structure and Feminine Status*. Paris: UNESCO, 1983.

———. "Work in the Lives of Caribbean Women." *Social and Economic Studies* 35, no. 2 (June 1986): 177–239.

Martin, J. et al. "Births: Final Data for 2003." *National Vital Statistics Report*. 54, no. 2 (2005): 1–116.

McClintock, Anne. *Imperial Leather: Race, Gender, and Sexuality in the Colonial Contest*. New York: Routledge, 1995.

Merry, Sally. "Transnational Human Rights and Local Activism: Mapping the Middle." *American Anthropologists* 108, no. 1 (2006): 38–51.

Mertus, Julie. "The Rejection of Human Rights Framings: The Case of LGBT Advocacy in the US." *Human Rights Quarterly* 29 (2007): 1036–1064.

Miller, Errol. *Men at Risk*. Kingston, Jamaica: Jamaica Publishing House, 1991.

———. "Male Marginalization Revisited Gender." In *Gender in the 21st Century: Caribbean Perspectives, Visions and Possibilities*, eds. Barbara Bailey and Elsa Leo-Rhynie, 99–133. Mona, Jamaica: Ian Randle Press, 2004.

Mohanty, Chandra. *Feminism without Borders: Decolonizing Theory, Practicing Solidarity*. Durham: Duke University Press, 2003.

Mohanty, Chandra et al. *Third World Women and the Politics of Feminism.* Bloomington: Indiana University Press, 1991.

Molyneux, Maxine. "Mobilization without Emancipation? Women's Interests, the State and Revolution in Nicaragua." *Feminist Studies* 11, no. 2 (1985): 227–54.

Morgan, Kenneth. *Slavery and the British Empire: From Africa to America.* London: Oxford University Press, 2008.

Morrissey, Marietta. *Women's Work, Family Formation and Reproduction among Caribbean Slaves.* East Lansing, MI: Michigan State University, 1984.

———. *Slave Women in the New World: Gender Stratification in the Caribbean.* Kansas: University Press of Kansas, 1989.

Morrow, Bruce, and Isaac Julien. "An Interview with Isaac Julien." *Callaloo* 18, no. 2 (Spring 1995): 406–415.

Moser, Caroline. Gender Planning in the Third World: Meeting Practical and Strategic Gender Needs." In *Gender and International Relations*, eds. Rebecca Grant and Kathleen Newland, 86–87. Bloomington: Indiana University Press, 1991.

———. *Gender Planning and Development: Theory, Practice, and Training.* London: Routledge, 1993.

———. "Evaluating Gender Impacts." *New Directions for Evaluation* 67 (Fall 1995): 105–118.

———. "'Has Gender Mainstreaming Failed?': A Comment on International Development Agency Experiences in the South." *International Feminist Journal of Politics* 7, no. 4 (December 2005): 576–590.

Murray, David. "Who's Right? Human Rights, Sexual Rights, and Social Change in Barbados." *Culture, Health and Sexuality* 8, no. 3 (May–June, 2006): 267–81.

Narayan, Uma. *Dislocating Cultures: Identities, Traditions, and Third World Feminism.* New York: Routledge, 1997.

Nunes, Fred, and Yvette M. Delph. "Making Abortion Law Reform Happen in Guyana: A Success Story." *Reproductive Health Matters: Women's Health Policies: Organizing for Change* 3, no. 6. (Nov. 1995): 12–23.

———. "Making Abortion Law Reform Work: Steps and Slips in Guyana" *Reproductive Health Matters* 5, no. 9. *Abortion: Unfinished Business.* (May 1997): 66–76.

Nussbaum, Martha C. *Women and Human Development: The Capabilities Approach.* Cambridge: Cambridge University Press, 2000.

Oakley, Ann. *Sex, Gender, and Society.* London: Temple Smith, 1972.

Obiora, L. Amede "'Supri, supri, supri, Oyibo': An Interrogation of Gender Mainstreaming Deficit. " *Signs: Journal of Women in Culture and Society* 29, no. 2 (2003): 649–662.

O'Flaherty, Michael, and John Fisher. "Sexual Orientation, Gender Identity and International Human Rights Law: Contextualizing the Yogyakarta Principles." *Human Rights Law Review* 8, no. 2 (2008): 207–248.

Parry, O. "In One Ear and Out the Other: Unmasking Masculinities in the Caribbean Classroom." *Sociological Research Online* 1, no. 2 (1996), http://www.socresonline.org.uk/1/2/2.html (accessed November 2008).

Pateman, Carole. "Equality, Difference, Subordination: The Politics of Motherhood and Women's Citizenship." In *Beyond Equality and Difference: Citizenship, Feminist Politics, and Female Subjectivity*, eds. Gisela Bock and Susan James, 17–31. New York: Routledge, 1992.

Payne, Anthony, and Paul Sutton. *Modern Caribbean Politics.* Baltimore: The John Hopkins University Press, 1993.

Peterson, V.S. *A Critical Rewriting of Global Political Economy: Integrating Reproductive, Productive, and Virtual Economies.* New York: Routledge, 2003.

Pharr, Suzanne. *Homophobia: Weapon of Sexism*. Arkansas: Chardon Press, 1988.
Port of Spain Gazeet. 27 September, 1826, n.p.
Puar, Jasbir Kaur. "Global Circuits: Transnational Sexualities in Trinidad." *Signs: Journal of Women in Culture and Society* vol. 26, no. 4 (Summer 2001): 1939–1066.
———. "Circuits of Queer Mobility: Tourism, Travel, and Globalization." *GLQ* 8, no. 1–2 (2002): 101–137.
Rai, Shirin. *The Gender Politics of Development: Essays in Hope and Despair*. London: Zed Books, 2008.
Ram, Kalpana. "Rationalizing Fecund Bodies: Family Planning Policy and the Modern Indian Nation-State." In *Borders of Being: Citizenship, Fertility and Sexuality in Asia and the Pacific*, eds. Margaret Anne Jolly and Kalpana Ram, 82–117. Ann Arbor: University of Michigan Press, 2001.
Rathgeber, Eva. "WID, WAD, GAD: Trends in Research and Practice." *Journal of Developing Areas* 24, no. 4 (July 1990): 489–502.
Reagon, Bernice Johnson. "Coalition Politics: Turning the Century." In *Home Girls: A Black Feminist Anthology*, ed. Barbara Smith, 356–368. New York: Kitchen Table Women of Color Press, 1983.
———. *Killing the Black Body: Race, Reproduction and the Meaning of Liberty*. New York: Vintage, 1997.
Reddock, Rhoda. "Historical and Contemporary Perspectives: The Case of Trinidad and Tobago." In *Women and the Sexual Division of Labour in the Caribbean*, ed. Keith Hart, 9–28. Kingston, Jamaica: Canoe Press, University of the West Indies, 1989.
———. "Women's Organizations and Movements in the Commonwealth Caribbean: The Response to the Global Economic Crisis in the 1980s." *Feminist Review* 59, no. 1 (Summer 1998): 57–73.
———. *Women, Labor and Politics in Trinidad and Tobago: A History*. London: Zed Books, 1994.
———. ed. *Ethnic Minorities in Caribbean Society*. Trinidad and Tobago: ISER, St. Augustine, 1996.
Rees, Teresa. "Reflections on the Uneven Development of Gender Mainstreaming in Europe." *International Feminist Journal of Politics* 7, no. 4 (December 2005): 555–574.
Roberts, Dorothy. *Killing the Black Body: Race, Reproduction, and the Meaning of Liberty*. New York: Pantheon Books, 1997.
Roberts, Dorothy. *Family Planning, Policy and Development Discourse in Trinidad and Tobago: A Case Study in Nationalism and Women's Equality*. Institute for Policy Research Working Paper Series WP-04-06. Chicago: Northwestern University Institute for Policy Research, 2003.
Robinson, Tracy. "Fictions of Citizenship, Bodies without Sex: The Production and Effacement of Gender in Law." *Small Axe* 7 (2000): 1–27.
———. "Authorized Sex: Same-Sex Sexuality and Law in the Caribbean." In *Sexuality, Social Exclusion and Human Rights*, eds. Christine Barrow, Marjan de Bruin, and Robert Carr, 3–22. Jamaica: Ian Randle Press, 2009.
Robinson, Tracy, Kalaycia Clarke, and Noelle-Nicole Walker. *Sexual Harassment and the Law in Barbados*. Prepared for the Coalition against All Forms of Sexual Harassment (CASH). Barbados: CASH, August 21, 2003.
Rohlehr, Gordon. *Calypso and Society in Pre-Independence Trinidad*. Port of Spain, Trinidad: Gordon Rohlehr, 1990.
Roseneil, Sasha. "The Global, the Local and the Personal: The Dynamics of a Social Movement in Postmodernity." In *Globalisation and Social Movements*, eds. P. Hamel, H. Lustiger-Thaler, J. Nederveen Pieterse, and S. Roseneil, 89–110. Basingstoke: Palgrave, 2001.

Rostow, W.W. *The Stages of Economic Growth: A Non-Communist Manifesto.* Cambridge, UK: Cambridge University Press, 1960.

Rowley, Michelle. "Feminist Visions for Women in a New Era: An Interview with Peggy Antrobus." *Feminist Studies* 33, no. 1 (2007): 64–87.

———. "When the Post-Colonial State Bureaucratizes Gender: Charting Trinidadian Women's Centrality within the Margins." In *Gender in the 21st Century: Perspectives, Visions and Possibilities*, eds. Barbara Bailey and Elsa Leo-Rhynie, 655–688. Mona, Jamaica: Ian Randle Press, 2004.

———. "Crafting Maternal Citizens? Public Discourses of the 'Maternal Scourge' in Social Welfare Policies and Services in Trinidad." *Social and Economic Studies* 52, no. 3 (2003): 31–58.

Roy, Sara. "De-development Revisited: Palestinian Economy and Society Since Oslo." *Journal of Palestine Studies* 28, no. 3 (1999): 64–82.

Rubin, Gayle. "The Traffic in Women: Notes on the 'Political Economy' of Sex." In *Toward an Anthropology of Women*, ed. Rayna Reiter, 157–210. New York, Monthly Review Press, 1975.

Sartre, Jean Paul. *Critique of Dialectical Reason, Volume I.* Translated by Alan Sheridan-Smith. New York: Vernon Press, 2004.

Saunders, Patricia. "Is Not Everything Good To Eat, Good to Talk: Sexual Economy and Dancehall Music in the Global Marketplace." *Small Axe* 13 (March 2003): 95–115.

Schaner, D.J. "Romance in the Workplace: Should Employers Act as Chaperones?" *Employee Relations Law Journal* 20, no. 1 (1994): 47–72.

Schultz, Vicki. "Telling Stories about Women and Work: Judicial Interpretations of Sex Segregation in the Workplace in Title VII Cases Raising the Lack of Interest Argument." *Harvard Law Review* 103, no. 8 (June 1990): 1749–1843.

Scott, Joan. "The Evidence of Experience." *Critical Inquiry* 17, no. 4 (Summer 1991): 773–797.

———. *Gender and the Politics of History.* New York: Columbia University Press, 1988.

Seebaran-Suite, Lynette. "Knowledge and Perception of Abortion and the Abortion Law in Trinidad and Tobago." *Reproductive Health Matters* 15, no. 29 (May 2007): 97–107.

Sedgwick, Eve. *Epistemology of the Closet.* Berkeley, California: University of California Press, 2008.

Sen, Amartya. *Development as Freedom.* New York: Anchor Press, 2000.

Sharpe, Jenny and Samantha Pinto. "The Sweetest Taboo: Studies of Caribbean Sexualities: A Review Essay." *Signs: Journal of Women in Culture and Society* 32, no. 1 (2006): 247–274.

Silvera, Makeda. "Man Royals and Sodomites: Some Thoughts on the Invisibility of Afro-Caribbean Lesbians." *Feminist Studies* 18, no.3 (Fall 1992): 521–582.

Silvestri, Marisa. *Women in Charge: Police, Gender, and Leadership.* Devon, UK: Willan Publishing, 2003.

Simey, T.S. *Welfare and Planning in the West Indies.* London: Oxford University Press, 1946.

Singh, Kelvin. "June 1937 Disturbances." In *The Trinidad Labour Riots of 1937: Perspectives 50 Years Later*, ed. Roy Thomas, 14–40. The University of the West Indies, St. Augustine, Trinidad: Extra Mural Studies Unit, 1987.

Singh, Susheela. "Hospital Admissions Resulting from Unsafe Abortion: Estimates from 13 Developing Countries." *Lancelet* 386 (2006); 1887–1892.

Sippel, Serra. "Achieving Global Sexual and Reproductive Health and Rights." *Human Rights* 35, no. 13 (2008): 1–4.

Smith, Raymond T. *The Negro Family in British Guiana: Family Structure and Social Status in the Villages*. London: Routledge and Kegan Paul Ltd., 1956.
———. "The Family in the Caribbean." In *Work and Family Life*, eds. L. Comilas and D. Lowenthal, 351–363. New York: Anchor Press, 1973.
———."Hierarchy and the Dual Marriage System in West Indian Society." In *Gender and Kinship: Essays toward a Unified Analysis*, eds. J. Collier and S.J. Yanagisako, 163–196. Stanford: Stanford University Press, 1987.
———. *Kinship and Class in the West Indies: A Genealogical Study of Jamaica and Guyana*. Cambridge: Cambridge University Press, 1988.
———. *Kinship and Class in the West Indies: A Genealogical Study of Jamaica and Guyana*. Cambridge: Cambridge University Press, 1990.
———. *The Matrifocal Family: Power, Pluralism, and Politics*. New York: Routledge, 1996.
Smith, Tony. "Requiem or New Agenda for Third World Studies?" *World Politics* 37, no. 4 (July 1985):532–561.
Spillers, Hortense. "Mama's Baby, Papa's Maybe: An American Grammar Book." *Diacritics*. (Summer 1987): 65–81.
Spivak, Gayatri Chakravorty. *A Critique of Postcolonial Reason: Toward A History of the Vanishing Present*. Cambridge, MA: Harvard University Press, 1999.
Standing, Guy. "Global Feminization Through Flexible Labor." *World Development* 17 no. 7 (1989): 1077–1095.
St. Hill, Donna. "Women and Difference in Caribbean Gender Theory: Notes toward a Strategic Universalist Feminism." In *Confronting Power, Theorizing Gender: Interdisciplinary Perspectives in the Caribbean*, ed. Eudine Barriteau, 46–74. Kingston, Jamaica: University of the West Indies Press, 2003.
Staudt, Kathleen ed. *Women, International Development, and Politics: The Bureaucratic Mire*. Philadelphia: Temple University Press, 1990.
Staudt, Kathleen. "Gender Mainstreaming: Conceptual Links to Institutional Machineries." In *Mainstreaming Gender Democratizing the State: Institutional Mechanisms for the Advancement of Women*, ed. Shirin M. Rai, 40–65. New Brunswick: Transaction Publishers, 2007.
Stean, Jill, and Vafa Ahmadi. "Negotiating the Politics of Gender and Rights: Some Reflections on the Status of Women's Human Rights at 'Beijing Plus Ten.' " *Global Society* 19, no. 3 (July 2005): 227–245.
Stetson, Dorothy McBride, and Amy Mazur, eds. *Comparative State Feminism*. Thousand Oaks, California: Sage Publications, 1995.
Stoler, Ann. *Race and the Education of Desire: Foucault's History of Sexuality and the Colonial Order of Things*. Durham: Duke University Press, 1995.
Stone-Mediatore, Shari. "Chandra Mohanty and the Revaluing of Experience." *Hypatia: A Journal of Feminist Philosophy* 13, no. 2 (Spring 1998): 116–133.
———. *Reading across Borders: Storytelling and Knowledges of Resistance*. New York: Palgrave Macmillan, 2003.
Strickland, L.A. "Female Police Perspective." Paper presented at 2005 Annual Force Conference, Royal Barbados Police Force Band Headquarters, Barbados, 2005.
Swarr, Amanda Lock, and Richa Nagar. "Dismantling Assumptions: Interrogating 'Lesbian' Struggles for Identity and Survival in India and South Africa." *Signs: Journal of Women in Culture and Society* 29, no. 2 (Winter 2004): 491–516.
Tamale, Sylvia "'The Personal Is Political,' or Why Women's Rights Are Indeed Human Rights: An African Perspective on International Feminism." *Human Rights Quarterly* 17, no. 4 (1995): 691–731.

Tambiah, Yasmin. "Creating Immoral Citizens: Gender Sexuality and Lawmaking in Trinidad and Tobago, 1986." *Caribbean Review of Gender Studies: A Journal of Caribbean Perspectives on Gender and Feminism* 3 (2009): 1–19. http://sta.uwi.edu/crgs/november2009/journals/Tambiah.pdf (accessed December 2009).

Thoreson, Ryan. "Queering of Human Rights: Yogyakarta Principles and the Norm That Dare Not Speak Its Name." *Journal of Human Rights* 8, no. 4 (2009): 323–339.

Tindigarukayo, Jimmy. "Perceptions and Reflections on Sexual Harassment in Jamaica." *Journal of International Women's Studies* 7, no. 4 (May 2006): 90–110.

Trotz, Alissa. "Guardians of Our Homes, Guards of Yours? Economic Crisis, Gender Stereotyping, and the Restructuring of the Private Security Industry in Georgetown, Guyana." In *Caribbean Portraits: Essays on Gender Ideologies and Identities*, ed. Christine Barrow, 28–54. Mona, Jamaica: Ian Randle Press, 1998.

True, Jacqui. "Mainstreaming Gender in Global Public Policy." *International Feminist Journal of Politics* 5, no. 3 (November 2003): 368–396.

Trinidad Guardian. "Royal Commission Hearings." 26 February, 1939, 12.

———. *Trinidad Guardian*. "Royal Commission Hearings" 4 January 1939, 1.

Underhill-Sem, Y. "Embodying Post-Development: Bodies in Places, Places in Bodies." *Development (Place, Politics, and Justice: Women Negotiating Globalization)*. 45, no. 1 (March 2002): 54–73.

United Nations. *Women 1980, Conference Booklet for the World Conference of the United Nations Decade for Women*, July 14–30, 1980. New York: UN Division for Social and Economic Information, 1980.

———. *The Nairobi Forward-Looking Strategies for The Advancement of Women: As Adopted by the World Conference to Review and Appraise the Achievements of the United Nations Decade for Women: Equality, Development and Peace*. Nairobi, July 15–26 1985. New York: UN Department of Public Information, 1986.

———. *The Beijing Declaration and the Platform for Action*. Fourth World Conference on Women, Beijing China, September 14–15 1995. New York: UN Department of Public Information, 1996.

———. "Report of the Economic and Social Council for 1997." *Official Records of the General Assembly, Fifty Second Session, Supplement No. 3*. UN: Department of Public Information, 1997. (A/52/3/Rev.1, chap. IV, para.4).

———. "Women and Health: Mainstreaming the Gender Perspective into the Health Sector," Expert Group Meeting, Tunis, Tunisia. September 28–October 2, 1998, United Nations Division for the Advancement of Women, Department of Economic and Social Affairs, United Nations. http://www.un.org/womenwatcNdaw/csw/healthr.htm. (accessed July 2007).

———. "United Nations Millennium Declaration," 2000 (Ref: A/55/L.2), www.un.org / millennium / declaration /ares552e.htm (accessed July 2001).

———. *The Human Rights-Based Approach to Development Cooperation: Towards a Common Understanding Among the UN Agencies (Statement of Common Understanding)*, 2003. http://www.undg.org/archive_docs/6959-The_Human_Rights_Based_Approach_to_Development_Cooperation_Towards_a_Common_Understanding_among_UN.pdf (accessed on June 5, 2010).

UNCEDAW. UN *Convention on the Elimination of All Forms of Discrimination against Women*, UN, 1979. (Doc. A/34/36).

United Nations Division for the Advancement of Women. "The Role of National Mechanisms in Promoting Gender Equality and the Empowerment of Women." Report of the Expert Group Meeting, Rome, Italy. November 29–December 2,

2004. New York: UNDAW, January 31, 2005. (EGM/National Machinery/2004/REPORT).

United Nations Development Fund for Women. "Institutional Mechanisms for the Advancement of Women in the Caribbean: Regional Assessment." Prepared by Michelle Rowley for the ECLAC/CDCC/UNIFEM/CIDA/CARICOM Fourth Caribbean Ministerial Conference on Women: Review and Appraisal of the Beijing Platform for Action, February 2–13, 2004, Saint Vincent and the Grenadines. http://www.iknowpolitics.org/files/InstitutionalMechanisms_Advancement_Caribbean.pdf (accessed June 10, 2010).

———. "Gender and the Millennium Development Goals: More Specific Targets and Indictors for the Caribbean." Report prepared for the Task Force on Gender and Poverty. Guyana: CARICOM Secretariat, June 22, 2005.

United Nations Economic Commission for Latin America and the Caribbean (UNECLAC). "Review of the Implementation of the Cairo Program of Action in the Caribbean (1994–2004)." Port of Spain: UNECLAC, October 14, 2003. (WP/2003/8).

Veitch, Janet. "Looking at Gender Mainstreaming in the UK Government." *International Feminist Journal of Politics* 7, no. 4 (December 2005): 631–638.

United Nations Population Information Network (UNPOPIN). "Newsletter of the International Conference on Population and Development Cairo, Egypt, 5–13 September 1994." ICPD 94, no. 14 (April 1994). http://www.un.org/popin/icpd/newslett/94_14/all14.html (accessed July 2007).

Visweswaran, Kamala. *Feminist Ethnography*. Minneapolis: University of Minnesota Press, 1994.

Vizenor, Gerald. *Manifest Manners: Postindian Warriors of Survivance*. Hanover, NH: Wesleyan University Press, 1994.

———. *Survivance: Narratives of Native Presence*. Nebraska: University of Nebraska Press, 2008.

Von Braunmühl, Claudia. "Mainstreaming Gender—A Critical Revision." In *Common Ground or Mutual Exclusion? Women's Movements and International Relations*, eds. Marianne Braig and Sonja Wolte, 55–79. London: Zed Books, 2002.

Wahab, Amar, and Dwaine Plaza. "Queerness in the Transnational Canadian-Caribbean Diaspora." *Caribbean Review of Gender Studies: A Journal of Caribbean Perspectives on Gender and Feminism* 3 (November 2009): 1–33, http://sta.uwi.edu/crgs/november2009/journals/Wahab-Plaza.pdf (accessed June 12, 2010).

Waites, Matthew. "Critique of Sexual Orientation and Gender Identity in Human Rights Discourse: Global Queer Politics beyond the Yogyarkarta Principles." *Contemporary Politics* 1, no. 15 (2009) 137–156.

Walby, Sylvia. "A Critique of Post Modernist Accounts of Gender," Paper presented at Canadian Sociological Association Meetings, Vancouver, British Columbia, 1990, quoted in *Feminism/Postmodernism/Development*, eds. Jane Parpart and Marianne Marchand, 5. London: Routledge, 1995.

Walcott, Rinaldo. "Outside in Black Studies: Reading from a Queer Place in the Diaspora." In *Black Queer Studies: A Critical Anthology*, eds. E. Patrick Johnson and Mae G. Henderson, 90–105. Durham: Duke University Press, 2005.

———. "Queer Returns: Human Rights, the Anglo-Caribbean and Diaspora Politics." *Caribbean Review of Gender Studies: A Journal of Caribbean Perspectives on Gender and Feminism* 3 (November 2009):1–19.http://sta.uwi.edu/crgs/november2009/journals/Walcott.pdf (accessed June 12, 2010).

Wallerstein, Immanuel. "After Developmentalism and Globalization, What?" *Social Forces* 83, no. 3 (2005): 1263–1278.

Weber, Max. *Economy and Society.* Translated by G. Roth and C. Wittich. Berkley, CA: University of California Press, 1978.
Weeks, Jeffrey. *Sexuality and Its Discontents: Meanings, Myths, and Modern Sexualities.* New York: Routledge, 1990.
Wegar, Katarina. "In Search of Bad Mothers: Social Constructions of Birth and Adoptive Motherhood." *Women's Studies International Forum* 20, no. 1 (1997): 77–86.
Wekker, Gloria. *The Politics of Passion: Women's Sexual Culture in the Afro-Surinamese Diaspora (Between Men–Between Women).* New York: Columbia University Press, 2006.
Wendoh, Senorina, and Tina Wallace. "Re-Thinking Gender Mainstreaming in African NGOs and Communities." *Gender and Development* 17, no. 2 (July 2005): 70–79.
Whaley-Eager, Paige. "From Population Control to Reproductive Rights: Understanding Normative Change in Global Population Policy (1965–1994)." *Global Society* 18, no. 2 (April 2004): 145–173.
Whiteley, Paul. "Economic Growth and Social Capital." *Political Studies* 48, no. 3 (February 2002): 443–466.
Wicks, David, and Patricia Bradshaw. "Investigating the Gendered Nature of Organizational Culture: Gendered Value Foundations That Reproduce Discrimination and Inhibit Organizational Change." In *Gender, Identity and the Culture of Organizations*, eds. Iris Aaltio and Albert Mills, 136–159. New York: Routledge, 2002.
Williams, Lawson. "Homophobia and Gay Rights Activism in Jamaica." *Small Axe* 7 (March 2000):106–11.
Wills, Magda. *Sexual and Gender Minorities Baseline: The Situation in Guyana, HIV Sexual Minorities Project, 2010.* http://www.sasod.org.gy/files/UNDP_Guyana_Sexual_and_Gender_Minorities_Report.pdf (accessed June 10, 2010).
WIRC (Moyne Report). *Report: West India Royal Commission 1938–1939.* Great Britain, 1945.
Yeboah, David. "A Demographic Profile." Paper presented at the Caribbean Studies Association Annual Meeting, St. Maarten/St. Martin, 26th, May 29–June 3, 2001.
———. "Strategies Adopted by Caribbean Family Planning Associations to Address the Challenges of Declining International Funding." *International Family Planning Perspectives* 28, no. 2 (June 2005). http://www.guttmacher.org/pubs/journals/2812202.html (accessed July 20, 2007).
Yelvington, Kevin. *Trinidad Ethnicity.* London: Macmillan Caribbean, Warwick University Caribbean Studies, 1993.
Young, Gary. "In Search of the Cartesian Self: An Examination of Disembodiment within 21st-Century Communication." *Theory and Psychology* 20, no. 2 (2010): 209–229.
Zippel, Kathrin. "Transnational Advocacy Networks and Policy Cycles in the European Union: The Case of Sexual Harassment." *Social Politics* 11, no. 1 (Spring 2004): 57–85.

Index

A

Aaltio, Iris, 129
abortion. *See* termination of pregnancy
Adams, John Michael Geoffrey Manningham ("Tom"), 119
Advocates for Safe Parenthood: Improving Reproductive Equity (ASPIRE), xxiv, 110–112, 122, 212n19
agency as resignification, 4, 197
Ahmadi, Vafa, 99
Ahmed, Sara, 190
Alexander, Jacqui, 45, 48, 91, 95, 183–186, 193, 206n2, 218n12
Ali, Shaheen, 5–6
Alston, Philip, 101–103, 105–106, 109
Alvarez, Sonia, 9, 12, 207n8
"ambivalent sexism," 159
Ambivalent Sexism Inventory, 158–159
Amelioration Period, 26–31, 32, 208n5
Amnesty International, 184
Anguilla: gender unit in 64; institutional mainstreaming in 210n1; within Anglophone Caribbean community 205–206n1
Annan, Kofi, 108, 214n17, 217n7
"anti-subordination approach" to sexual harassment, 151
Antigua and Barbuda: within Anglophone Caribbean community 205–206n1; women's reproductive health in 113, 114, 214n25
Antrobus, Peggy, 62–63, 103–104, 107, 108
Appadurai, Arjun, 44
Apprenticeship Period, 26, 208n3
Armas, Henry, 176–177
Atilla, 208n12

B

Bacchi, Carol, 56–57, 127, 207n11
"bad sex," 176. *See also* Jolly, Susan
Baden, Sally, 17–18, 207n11, 207n12
Bahamas: domestic violence legislation in, 216n8; sexual harassment legislation in, 154, 155, 216n8; within Anglophone Caribbean community, 205–206n1; women's reproductive health in, 113
Bailey, Barbara, 140, 148
Bakhtin, Mikhail, 13, 81
Banton, Buju, 195–196
Barbados: GDI, GEM, and HDI ranking of, 70; gay rights (*see* sexuality/rights in); gender unit in, 64, 65; institutional mainstreaming in, 210n1; sexual harassment and gender equality in, xxiii, xxv, 153, 154, 156, 161, 163, 164, 216 n7 (*see also* CASH, RBPF); sexuality/rights in, 189, 219n22; within Anglophone Caribbean community, 205–206n1; women's reproductive health/rights in, xxv, 92, 95, 113, 114, 116, 117, 119, 120–121, 122, 143, 144, 214n26, 215n33, 215n34
Barbados Workers' Union, 161, 215n34
Barriteau, Eudine, 66, 79, 81, 91, 211n16
Basu, Amrita, 10
Baugh, Edward, xxi–xxii, 214n24
Beall, Jo, 6, 59, 91, 207 n6
"because of sex," 151–152. *See also* Franke, Katherine
"Before Difference/After Difference," 128

238 *Index*

Beijing Declaration and Platform for Action, xxii, xxiii, 6, 8, 9, 59, 71–72, 79, 92, 95, 98–100, 106, 111, 116, 125, 181, 206n3, 217n7
Belize: gender unit in, 65; institutional mainstreaming in, 210n1; sexual harassment legislation in, 65, 154, 155, 216n8, 216n10; within Anglophone Caribbean community, 205–206n1; women's reproductive health in, 113
"benevolent sexism," 159
Bentham, Jeremy, 46
Berkovitch, Nitza, 120
Berlant, Lauren, 171, 172, 203, 204
Bhabha, Homi, 26
bisexual, 217n3
Black Power Movement, 72
Binnie, Jon, 200
black m(O)ther. *See* M(O)therism
Bordo, Susan, 3, 212n2
Boserup, Ester, 21, 91, 207–208n1
Bradshaw, Patricia, 129
Braidotti, Rosi, 93, 94, 204
Brereton, Bridget, 208n7
British Virgin Islands: gender unit in, 64; institutional mainstreaming in, 210n1; within Anglophone Caribbean community, 205–206n1
Brown-Campbell, Gladys, 164–165
bullerman, 194, 218n16
bureaucracy (bureaucratic cultures/systems), xxiv, 8, 10, 12, 14, 58–59, 61–62, 66–67, 71, 74, 76–78, 80, 85, 87, 120, 123, 125, 144, 153, 165, 167, 207n8
"but for sex," 151–152. *See also* Franke, Katherine
Butler, Judith, 4, 16, 33, 171, 190, 217n6
Butler, Uriah, 33

C

Cairo Conference. *See* International Conference on Population and Development
calypso, 171, 194, 208n12
Calypso Rose. *See* McCartha Lewis
Canadian International Development Agency (CIDA), 9, 108, 210n6, 211n10

Caribbean Forum for Liberation and Acceptance of Gays (CARIFLAGS), 219n22
Caribbean Association for Feminist Research and Action (CAFRA), 73, 211n10, 212n19
Caribbean Community (CARICOM) 118, 155, 206n1, 211n12, 213n9, 213n11, 216n10
Caribbean Development Bank (CDB), 102, 107, 213n11
Carney, Gemma, 6, 12–13, 14
cartography, 1, 94, 186, 198, 203–204
Chevannes, Barry, 42, 140, 192–193
Chin, Timothy, 193, 195
"Christian Family," 39
Choice: regarding abortion, 97, 100–101, 110–111, 125–126; regarding profession, 147–148
Chowdhury, Elora, 9–10
Citizenship, 120. *See also* reproductive citizenship; sexual citizenship
clientilism, 11, 86, 207n9. *See also* donor agencies
Coalition Advocating for Inclusion of Sexual Orientation (CAISO), 219n22
coalition-building, xxv, 13, 18–19, 59, 63, 68, 111, 119. *See also* CASH
Coalition against All Forms of Sexual Harassment (CASH), xxv, 153–154, 156, 215n1
coincidence versus causality, 8–9
Colonial Development and Welfare Office (Trinidad), 38, 40
"colonial erotic," 184
common-law relationship rights, 74
"Compassionate Family," 39. *See also* Simey, T.S.
compensatory agenda, 153
Conference for Women in the Caribbean, 63, 210n1
conferences. *See* CARICOM; Conference for Women in the Caribbean; UN Conferences for Women
contextuality and contingency of gender, 4, 11, 18, 22–23, 585, 69, 91–92, 94, 97, 114–115, 119, 173. *See also* the "local"
contraception, 110, 111, 112, 214n19, 214–215n27
Convention for the Elimination of All Forms of Discrimination Against

Women (CEDAW), 5, 8, 71–72, 74, 80, 98, 106, 108, 123, 125, 180, 181, 211n12, 211n13, 218n11
Cook, Rebecca, 96–97
Coterie of Social Workers (Trinidad), 33
Crichlow, Wesley, 194, 198, 218n16
cross-dressing, 203
Cuba: gender unit in, 64, 66; institutional mainstreaming in, 210n1; women's reproductive health in, 113
cultural authenticity, 184, 196
culture: as rationale for "norms"/treatment of sexual minorities 192–193, 195; for treatment of women 163–164
cycle of certification, 84

D

decriminalization of abortion, 66, 96–101, 112, 114–118, 215n33. *See also* termination of pregnancy
Delph, Yvette, 115, 116, 118, 122
Department of Social Services (Trinidad), 46
Development, 67–72; and gender equity in the Caribbean, xxiii, 68, 69–72; and gender mainstreaming, 6–7, and reproductive rights, 125–126; and sexual minorities, xxv; and women, xxiii, xxv; as discourse, 2–3 (*see also* regimes of representation); critique of, 2, 125–126 (*see also* post-structuralism; transnational feminism); ethics of, xxv, 104–105; institutionalization of, 2–3, 7–8 (*see also* welfare services); professionalization of, 2–3 (*see also* women's bureaus/desks/units; Trinidad GAD); women's marginalization in, 21
"development as discourse," 2–4. *See also* Escobar, Arturo
Development Alternatives with Women for a New Era (DAWN), 10, 62, 103, 123
"developmentalism," xxiii, xxv
Dickens, Bernard, 96–97
"difference distraction," 128, 132–134, 139–140, 150, 151, 160

"Disintegrated Family," 39. *See also* Simey, T.S.
"dismembered citizenship," 108
domestic violence: 4, 50, 62, 173, 176; legislation in the Bahamas, 155, 216n8; legislation in Barbados, 65; legislation in Dominica, 66; legislation in St. Vincent and the Grenadines, 111; legislation in Trinidad and Tobago, 73, 79, 84
Dominica: gender unit in, 64, 66; heterosexualization of masculinity in, 192; institutional mainstreaming in, 210n1; within Anglophone Caribbean community, 205–206n1; women's reproductive health in, 113
donor agencies, 7, 9, 10, 17, 57, 58, 59, 65, 210n6
Douglas, Mary, 93–94
"down-low," 187–188
Drag, 188
Dunlop, Joan, 99

E

education, access to, 68, 69, 73, 86, 90, 103, 104, 105, 107, 108, 181, 183, 212 n7. *See also* skills training programs
Eisenstein, Hester, xxiv, 58, 91, 124, 205n4, 210n3
El-Bushra, Judy, 8, 17, 18, 207n13
embodied difference, 128, 141, 150. *See also* gender in/equity
embodied equity, 91–95, 98, 101, 104, 123, 125–126, 172–175, 183, 204; and the constitution/nation, xxii; and queer bodies, 172–181, 183, 202–203; and women's bodies, xxiv–xxvi, 104–105. *See also* reproductive rights; sexual harassment; sexuality rights
embodied sexuality, 175, 188
"embodied spatialities," 11, 188, 202–203
embodiment, 93–94
employment legislation, 66, 143, 216n8. *See also* "mainstreamed" in atypical professions
Enloe, Cynthia, 185
equity: approaches to, xxi, xxv–xxvi, 18, 21-; and embodiment, 94; and gender mainstreaming, xxii–xxiv, 1–19; and employment,

127; and reproductive rights, xxiv–xxv, 98–101, 109–110; and sexual harassment, xxv; and sexuality rights, xxv–xxvi; in/effectiveness in achieving gender equity, xxiii, xxiv–xxvi; through practice of coalition, 18–19; versus equality, 5–6, 98; versus identity, 18–19, 20. *See also* "mainstreamed" into atypical professions; reproductive rights; sexual harassment; sexuality rights

Escobar, Arturo, 2–3, 4, 20, 205n1
"ethics of development," xxv, 104–105
Eveline, Joan, 56–57, 207n11

F

"Faithful Concubinage," 39. *See also* Simey, T.S.
family form(ation)s, 36, 37, 39, 40–41, 42, 43, 47
family planning: approach to reproductive rights, 110–113, 116, 119, 120, 125; as right, 99, 108; associations, 108, 108, 110, 116, 117, 120, 215n33, 215n34
female: ascendancy 79; inability 134
feminist development practitioners, 7, 56–62. *See also* Trinidad GAD
feminist NGOs, 9, 10, 65, 79, 88, 111. *See also* ASPIRE, CASH, DAWN
feminists: liberal, 3; Marxist, 3; second wave, 73
"femocrat," xxiv, 55, 58–59, 72–86, 124, 125, 205n4. *See also* Hester Eisenstein
"fictions of feminine identity," 45
Fiske, Susan, 131, 158–159
Flax, Jane, 1, 19
Flood-Beaubrun, Sarah, 124
Foucault, Michel, 22–23, 25, 35, 45–36, 174, 192, 195, 199
Franke, Katherine, 151–152, 160, 168
Frazier, E. Franklyn, 37, 39–40
Friedman, Elizabeth, 11–12

G

gay: identity, 174, 184, 189, 190, 191, 217n3; in Jamaica, 186–187. *See also* Men who have Sex with Men (MSM)
gay rights: platform, rejection of by queer subjects, 189, 191–192; in Barbados, 189; in Guyana *See* SASOD. *See also* sexuality rights
gender: additive understanding, of 7, 14, 17, 75, 80, 81, 206n4; as analytical category, 13, 53, 178, 180, 192; as conceptual management tool, 24–25; conceptualizations of, xxiii, xxiv, xxvi, 6–7, 14–18, 21–24, 53–54, 56–57, 89–90, 172–174; decentering of, xxii, 57; "dilemma," 78–81; historicization of, xxiii, 20–23, 54, 60–61, 129 (*see also* gender-as-genealogy); illegality, xxiv, 74–78; institutionalization of 20–21, 24–44, 54, 55–62, institutionalization of in the Caribbean, 62–72; institutionalization of in Trinidad and Tobago, 35–53, 72–90 (*see also* T.S. Simey; R.T. Smith); in organizational culture, 61–62, 128–129 (*see also* RBPF); sensitization/training, 12, 13, 62, 63–64, 66, 82, 84, 114, shift from "woman" to, 75, 78–79, 85–86; understanding of, xxiii, 4, 7, 16, 57, 75, 81; versus "sex," 15–16, 17
"Gender as a Second Language," 13
"gender dilemma," 78–79, 81–82
gender equality, xxiv, xxviii, 9, 21, 55–57, 72–74, 76, 95, 97–98, 100, 105–107, 109, 127, 151, 212n7
gender equity, xxii–xxv, 1–14, 18, 20–21, 24, 25, 44, 53, 55, 56, 59, 64, 68–72, 74, 81, 86, 87, 90, 91–94, 97, 109, 125, 127–128, 130, 131, 133, 150, 153, 157, 196, 206–207n5, 215n1, 216 n7. *See also* embodied equity; gender mainstreaming; reproductive rights; sexuality rights
Gender and Development (GAD), 64, 67, 79, 88–90
"gender-as-genealogy," xxiii, 14, 17, 22–25, 26, 35–44, 204
Gender-related Development Index (GDI), 9, 59, 68–70; for Barbados, 70; for Grenada, 70; for Jamaica, 70; for Trinidad and Tobago, 70

Gender Empowerment Measure (GEM), 9, 59, 68–71, 106; for Barbados, 70; for Grenada, 70; for Jamaica, 70; for Trinidad and Tobago, 70
"gender expert," 12–13
gender focal point, 55, 63–65, 74–78, 82; definition of, 63. See also Trinidad GAD; women's bureaus/desk/units
gender formations. See gender-as-genealogy
gender harassment, 152
gender identity and sexual orientation, 190
gender in/equity, 18, 129. See also "mainstreamed" in atypical professions
gender-justice, xxvi, 8–9, 59–60, 63–64, 68, 77, 78, 84, 87, 88, 103
gender mainstreaming: and institutional frameworks, 56–62; approaches to 59–60; approaches to in St. Lucia, xxiv–xxv; approaches to in Trinidad and Tobago, xxiii–xxiv; as a form of state feminism, 8, 56, 77, 90, 210n3; as a mode of equity, 62, 127, 128; strategies for (see Beijing Platform; Cairo/ICPD; CEDAW; MDGs; Nairobi FLS; NGOs; UNHRC); constraints to, 56–62, 101–104, 109–110; constraints to in the Caribbean 62–72, 125–126; definitions of, 6–7; in the Caribbean, 62–72; in Trinidad and Tobago, 72–90; international overview of, 56–62; and reproductive rights, 91, 95; transformative possibilities of, 8; UN definition of, 7. See also gender equity; reproductive rights; sexuality rights
Glick, Peter, 131, 158–159
"global gay," 200, 219n21
Global South, xxiii, 5, 8, 9, 14, 21, 57, 58, 176, 178, 200, 205 n1, 206n2, 210n5, 218n11
"globalized developmentalism," 102, 213n8
Goetz, Ann-Marie, 8, 17, 18, 56, 62, 67, 91, 207n11, 207n12, 210n7
"good police officer," 135
Gosine, Andil, 174, 191, 193, 194

Grenada: GDI, GEM, and HDI ranking of, 70; within Anglophone Caribbean community, 205–206n1; women's reproductive health in, 113
grief as resource for politics, 171
Gutzmore, Cecil, 193–194, 219n19
Guyana: equal opportunity legislation in, 154, 216n8; gay rights (see sexuality/rights in); heterosexualization of masculinity in, 192; sexual harassment and gender equality in, 135, 141–142, 155, 156, 216 n8, 216n11; sexual violence legislation in, 219n28; sexuality/rights in, xxiii, xxv, 172, 198, 199, 201, 202 (see also SASOD); within Anglophone Caribbean community, 205–206n1; women's reproductive health in, xxiii, xxv, 92, 95, 113, 114–115, 118, 119, 120–121, 122, 214n22, 214n27
Guyana Rainbow Foundation (GUYBOW), 219n22

H

Haiti, women's reproductive health in, 113
Halberstam, Judith, 173, 175
Harcourt, Wendy, 4
Harris, Sonja, 67, 211n16, 212n18
Hayes, Ceri, 106
head of household, 47
heteronormativity, 61, 107, 114, 159, 174, 178, 183, 184, 185, 188, 190, 194, 204; and citizenship, 204; in development, 174–176; in human rights frameworks, 190; in sexual harassment conceptualizations, 158–159; in state governance, 174, 183, 185
heteropatriarchal resistance, 56, 68, 78, 100, 120, 128. See also masculinized organizational cultures
"heterosexual matrix," 190
Heyzer, Noeleen, 108
"higglers," 69
Higman, Barry, 26, 28, 29–30, 208n8
HIV/AIDS, 57, 102–103, 108–109, 112, 212n6, 214–215n27; and development, 176; and human rights activism, 180–181; coupling Caribbean with, 194,

218n18; coupling gay rights activism with, 180–181, 189, 191; coupling reproductive rights with, 121, 125; coupling sexual minority rights with, 191, 194; in the Caribbean 194, 218n18

Holland, Janet, 175

Homophobia: as nationalism, 184–185, 194; as resistance to neo-imperialism, 184–185; in the Caribbean (over-representation of), 172–173, 184–187, 191, 192–197, 201, 203, 219n20; in the diaspora, 171, 180–181, 186–188, 195–196; state-sanctioned, 184–185. *See also* heteronormativity; homosexual fetish; variegated hostilities

"homosexual fetish," xxv, 185

homosexual/heterosexual binary, 173

Housewife Association of Trinidad and Tobago, 73

"housewifization," 208n4

Human Development Index (HDI), xxiii, 68–70; for Barbados, 70; for Grenada, 70; for Jamaica, 70; for Trinidad and Tobago, 70

Human Development Report (HDR), 59, 68–70, 113

human rights framework, 89, 106, 177, 180, 218n11. *See also* reproductive rights; sexuality rights; women's rights as human rights

I

IDB Non-Traditional Skills Training Program, 74, 82, 85, 88, 211n17

inauthentic/alien (feminism as), 59, 65, 77, 97, 162, 182, 184

"incarcerated native," 44

inclusion, terms of, 68, 78, 85, 153, 203

"incompetence," 74, 76

"indigenization," 218n11

Inter-American Development Bank (IDB), 74, 82, 211 n17. *See also* IDB Non-Traditional Skills Training Program

International Conference on Population and Development (ICPD), 92, 95, 98–99, 107, 108, 116, 212n3

intergovernmental organizations (IGOs), 10, 160

International Lesbian and Gay Association (ILGA), 179, 217n9

International Monetary Fund (IMF), 72

J

Jahan, Rounaq, 60, 210n6

Jamaica: GDI, GEM, and HDI ranking of, 70; gay rights (*see* sexuality/rights in); gender unit in, 62, 64, 66; heterosexualization of masculinity in, 192; institutional mainstreaming in, 210n1, 210n7; sexual harassment and gender equality in, 157, 164; sexuality/rights in, 187, 195, 196, 219n19, 219n20, 219n22; slavery in, 29; within Anglophone Caribbean community, 205–206n1; women's reproductive health/rights in, 102, 111, 113, 124.

Jamaica Forum for Lesbians, All Sexualities and Gays (J-FLAG), 219n22

Jolly, Susan, 176

Julien, Isaac, 195–196

K

Kabeer, Naila, 5, 78, 104, 108

Kardam, Nuket, 8, 65, 68

Kempadoo, Kamala, 218n14

L

Lazarus-Black, Mindie, 50, 209nn25–26

Lazreg, Marnia, 24

lesbian: baiting, 157, 159, 161, 167–168; identity, 174, 175, 178, 179, 190, 193, 194, 217 n3; sexuality, 175, 218n12

Lewis, McCartha, 171

Lesbian, Gay, Bisexual, Transgender (LGBT): activism 179, 189, 195, 199–200; as identity/organizing terms, 190–191, 197, 217 n3; discrimination against, 174, 194, 196–197, 200, 201; in the Caribbean, 191, 197, 199, 219n22 (*see also* SASOD)

Lloyd, Moya, 189

"local," the, xxiii, 4, 8, 9, 10–12, 15, 16, 20, 21, 22, 24–25, 37, 53–54, 58, 69, 92, 140, 182, 187, 200, 202–203, 218n11

Lorde, Audre, 218n16

M

MacKinnon, Catherine, 141, 146, 150–151, 165
"madness," xxiv, 76–77
"mainstreamed" into atypical professions, 84, 127, 128, 132, 150, 153, 157, 170
male marginalization/reverse discrimination, xxv, 22, 41–43, 63, 75, 80–81, 211n16
Male Support Unit (Trinidad), 80–81
"malestream," 62
"man-royal," 194
Manning, Patrick, 123
masculinized organizational cultures, 128, 141, 146, 157, 158, 159, 168, 170
maternal: body, 27, 32, 40, 44–54, 126; centrality, 25, 39, 42, 43; "dysfunction," 25, 28, 32–35, 54; "family," 39–40; "fear," 131; ideal, 29; institutionalized disciplinization of, 25, 35–54; mortality in Trinidad, 111–112, 122–123; mortality worldwide, 100–101; sexuality, 26, 48–49; subjectivity, 30, 34, 37, 40, 47, 49, 51; "symbolic," 25, 32; trope, xxiv. See also matrifocality; maternal identity; m(O)therism; RBPF, motherhood and policing in
maternal identity: and citizenship, 55–90; and colonialism, xxiii–xxxiv; and nationalism, xxiii–xxiv; and slavery, 28, 30; and post-slavery, 39; and MDGs, 106; in Trinidad, xxiii–xxiv, 25–31, 35–44, 47–52; woman's identity as, 107–108
maternity leave: in Barbados, 143; in Jamaica, 111; in RBPF, 145
matism, 190, 218n15
"matrifocal complex," 42
matrifocality, 29, 40–43
maximum leadership, 8, 67. See also People Power
McClintock, Anne, 185
men who have sex with men (MSM), 187, 191, 218n18
Merry, Sally, 182, 218n11
Mertus, Julie, 179, 217–218n9

Millennium Development Goals (MDGs), 92, 95, 101–110, 111, 122, 125, 212–213n7, 213n9, 213n13, 214n20
Miller, Dame Billie, 119–120, 124, 215n32
Miller, Errol, 75, 211n16
Mills, Albert, 129
Modernity: development and, 68–72, 101–102 (see also CEDAW, GDI, GEM, HDI, HDR, MDGs, UN Conferences); ruse of, 200–201
Mohammed, Patricia, 89–90
Mohanty, Chandra, 19, 199
Moser, Carolyn, 7, 57, 60, 208n1
m(O)therism, 25–29, 30, 31, 54
Moyne Commission. See WIRC
Murray, David, 189, 191, 194

N

Nagar, Richa, 174, 178–179
Nairobi Forward-Looking Strategies, 7, 72, 88, 178
Narayan, Uma, 97
National Gender Policy for Trinidad and Tobago, 89–90
national machinery, xxiv–xxv, 6, 8, 9, 23, 46, 58, 62, 67, 68, 72, 77, 78, 81, 110, 205n3, 205n4, 207–208n1
nationalism: and development/modernity 68 (see also GDI, GEM, and HDI); and gender, 31–37, 129 (see also maternal; matrifocality); and reproductive rights, 112; and "rights" xxii, 182; and rights of sexual minorities, 171; in Trinidad, 45
"natural increase," 27, 29, 30
"negotiated continuity," 32
(neo-)liberalism, xxiii, 5, 17, 123, 174, 177, 217n4
non-governmental organizations (NGOs), 7, 9, 10, 57–58, 65, 76, 77–78, 79, 84, 88, 98, 111–112, 174, 178, 198, 215n32, 217n9. See also ASPIRE, CASH, DAWN, SASOD
Nunes, Fred, 115, 116, 118, 122, 215n29

O

Oakley, Ann, 15
Obiora, L. Amede, 57
O'Flaherty, Michael, 180

244 *Index*

Order of 1824 (Trinidad), 30–31
Organization of American States (OAS), 211n10
Organization of Eastern Caribbean States, 66
organizational culture, gendered, 128–129. *See also* masculinized organizational cultures

P

Pan African Movement, 32
Pan-American Health Organization (PAHO), 109, 211n10, 214n20
panoptic/ism, 44–49
panopticon, 46
paramilitary organizations. *See* RBPF
"participatory demagoguery," 67
"participatory democracy," 67
paternalism, 41, 42, 57, 100, 114, 116, 130, 158, 159
"People Power," xxi–xxii, 203
personhood, 120
Pharr, Suzanne, 167
Pinto, Samantha, 192–193
Plaza, Dwaine, 187
"policy evaporation," 60. *See also* Caroline Moser
"policy heteroglossia," 13–14, 81–82
policy planning: gender in, 55–62; in the Caribbean, 62–72; in Trinidad and Tobago, 72–90. *See also* gender focal points; women's bureaus/desks/units
political activity, 70
political correctness ("doing something for women"), 86
political commitment/will, 1, 8, 9, 44, 47, 57, 58, 63–64, 66–68, 85, 86–91, 123, 124
positionality, 16, 31, 197, 202, 207n10, 217n3
post-development, xxiii, 2. *See also* post-structuralism
post-structuralism, xxiii, 2–5
"potential difference," 142
"problem of family organization," 38
"problem of social readjustment," 38
"producing proof," 50–52
Protector of Slaves (Trinidad), 31
Puar, Jasbir, 188, 193

Q

queer rights as human rights. *See* gay rights; sexuality rights

queer subjectivity: and place, 176–177, 178–179, 184–186, 187, 194, 200–203, 204; and (post/colonial) time, 183–186, 201, 204; in Barbados, 189, 219n22; in Guyana (*see* SASOD); in Jamaica, 187, 219n19, 219n22; in Trinidad, 188, 194–195, 218n12, 219n22; versus LGBT identity politics 172, 173, 175, 180, 187, 189–191, 204 (*see also* homosexual/heterosexual binary)

R

"race for ratification," 55, 71–74, 211n12
Rai, Shirin, 206n2
Ram, Kalpana, 110
rape, 176, 201, 219n28; and termination of pregnancy, 113, 114, 115, 124, 214n23, 214–215n27. *See also* sexual violence, protection from
Rape Crisis Society (Trinidad), 73
Reagon, Bernice Johnson, 19
Reddock, Rhoda, 26, 73, 111, 206n1, 208n4, 208n15, 210n2
Rees, Teresa, 56
"regime of representation," 2, 4, 20, 22
re/membering the body, 92, 94, 95, 174. *See also* embodied equity
reproductive autonomy, 96, 991, 111, 121, 125
reproductive citizenship, 98–101, 125–126
reproductive health, xxii, xxvi, 65, 94, 98–100, 102–103, 104, 106–107, 108, 111–113, 114, 121–123, 125, 142, 212n4, 213n12, 214n16, 214n18. *See also* reproductive rights; termination of pregnancy
reproductive rights, xxvi, 91–126; and MDGs, 101–110; as human rights, 96, 98, 99, 106; in Caribbean, 91–101, 104, 107–108, 109, 110–125; in Barbados, xxiii, xxv, 92, 113, 141–146; in Guyana, xxiii, xxv; in Jamaica, 102, 111, 113; in St. Lucia, xxiii, xxiv, 66, 92, 95, 113; in Trinidad, xxiii, xxiv, 92, 95, 110, 113; in the workplace, 141–146. *See*

also ASPIRE; family planning, associations; termination of pregnancy
resources and/versus rights, 1, 6, 57
Rienzi, Adrian Cola, 33
"right to do and be," 186
"rite of humiliation," 50. *See also,* Lazarus-Black, Mindie
Roberts, Dorothy, 40, 110, 112, 115, 214n19, 214n21
Robinson, Tracy, 154, 156, 184, 215n30, 215n31
Royal Barbados Police Force (RBPF), xxv, 127–150, 160–170, 216n7; gendered organizational culture of, 129–131, 139, 146–150; defining sexual harassment in, 161–164; the maternal/motherhood and policing in, 130, 131, 141–146; placement in, 149; policing as gendered in, 135–141; reporting/responding to sexual harassment in, 164–170; sexual harassment in, 160–170; women's hypervisibility and culpability in, 132–134, 139, 167
Rubin, Gayle, 15

S

Safe Motherhood Conference, 213n13
same-sex unions, 90
Sandy, Michael, 171, 186, 187
Sartre, Jean Paul, 183
Saunders, Patricia, 193, 196
Schultz, Vicki, 147–148
"scientific method," anthropological, 37–38
Scott, Joan, 15–16, 20, 53, 72
Sedgwick, Eve, 173
self-surveillance, 46, 51, 132, 187, 199, 210n27
Sen, Amartya, 17, 105, 112
"sex" versus "gender," 15–17, 129, 151
sexual agency, 26, 48, 122, 160, 167, 183, 197, 199
sexual autonomy, xxv, 73, 91, 95–96, 183
sexual citizenship: defining, 171–172; and un/belonging, 174, 192, 197–198, 200, 203–204; transnational, 199, 200, 203. *See also* sexual minorities; sexuality rights
sexual harassment: "ambivalent ," 158–160; as anti-subordination, 151; as gender harassment, 151–152; as hostile working environment, 154; as sex discrimination, 154; as technology of sexism, 152, 168; as women's systematic vulnerability, 151; conceptualizations of ,150–152, 157–160, 162–164, 168 (*see also* gender harassment); dailyness and everydayness of, 163, 165–166; female-on-male, 152, 168; in atypical professions, 157–160 (*see also* "mainstreamed" in atypical professions); quid pro quo, 154, 161, 165, 168; same-sex 152, 168, 169; in Barbados, xxiii, xxv, xxvi, 160–170 (*see also* RBPF); in the Caribbean 161, 164
sexual harassment legislation: as fairness in the workplace, 160; in the Caribbean, 153–157; in the Bahamas, 154, 155, 216n8; in Barbados, 153–154, 216n7 (*see also* CASH); in Belize, 154, 155; in Guyana, 154, 155, 216 n8, 216n11; in Jamaica, 157; in St. Lucia, 66, 155, 216n8; in Trinidad, 154, 155, 216nn8–10
sexual minorities: and development, xxii, xxv, 172, 174–181; and government frameworks, 1, 72, 174–175, 177, 191, 197, 198–200, 203; rights for, xxiii, xxvi (*see also* sexuality rights as human rights); and social life, 172, 174, 175, 186, 198, 200, 203; hypervisibility and invisibility of, 191, 192–197, 202, 219n21; in Guyana, xxv (*see also* SASOD). *See also* LGBT
sexual orientation discrimination: in development ,174, 179; in Barbados, 189; in Guyana, 199–201; legislation in Guyana, 199–200, 201–202, in Trinidad, 90; UN and, 178–180 (*see also* Yogyakarta Principles
sexual pleasure (as human/sexual right), 99, 174–176, 183, 191, 204
sexual progressiveness, 180
"sexual survivance," 172, 197–198, 203. *See also* SASOD
sexual violence: protection from, 111, 113, 155, 201, 216; systematic terror by, 92, 212n1

sexual well-being, 125, 213n12
sexuality: and development, 172, 173, 174–181; as category of analysis 177–179, 192; de-essentializing of 173–175; defining 172–174; grounded understanding of, 194; in the Caribbean xxv, 172–173; integrality and indivisibility of within development, 178
sexuality rights: activism in the Caribbean, 219n22 (*see also* SASOD); and transnational activism, 199, 200, 202–203; as human rights, 171–172, 176–179, 181, 189–190, 198–199, 203, 218n10; in development, 172–181. See also SASOD; Stamford Statement of Common Understanding; UN, Declaration of Human Rights 189; UN, HRC Statement on Human Rights, Sexual Orientation and Gender Identity; Yogyakarta Principles
Sharpe, Jenny, 192–193
Silvera, Makeda, 194
Simey, T.S., 37–40, 209n17, 209n18
Singh, Susheela, 101, 212n6
skills training programs, 81–82. *See also* IDB Non-Traditional Skills Training Program
slavery: and black motherhood 27–31; and family formations, 31; in Jamaica, 29; in Trinidad and Tobago, 25–31
Smith, R.T., 37, 40–43
Smith, Tony, 205n1
Society Against Sexual Orientation Discrimination (SASOD), xxv, 172, 197–203, 219n22, 219n24–28, 220n32
sodomite, 194
"speculative vernaculzarization," 182, 189, 197. *See also* Yogyakarta Principles
Spivak, Gayatri, 172
"spoiling the man," 50
St. Kitts and Nevis: gender unit in, 64, 66; institutional mainstreaming in, 210n1; within Anglophone Caribbean community, 205–206n1; women's reproductive health in, 113, 124
St. Lucia: equal opportunity legislation in, 156, 216n8; gender unit in, 64, 66; institutional mainstreaming in, 210n1; sexual harassment legislation in, 155, 216n8; sexual harassment and gender equality in, 156; within Anglophone Caribbean community, 205–206n1; women's reproductive health in, xxiii, xxiv, 92, 95, 113, 124
St. Vincent and the Grenadines: campaign against violence to women in, 111; gender unit in, 64; institutional mainstreaming in, 210n1; within Anglophone Caribbean community, 205–206n1; women's reproductive health in, 113
Stamford Statement of Common Understanding, 177–178
state feminism, 8, 56, 77, 90, 210n3
state-managers, xxiv, 1, 7, 9, 19, 55, 71, 73, 75, 76, 79, 90, 91, 104, 109, 111, 112, 123, 174, 181, 183, 184, 185, 206n2
state formation, 25, 47, 79, 206n2, 208n6
Statement on Human Rights, Sexual Orientation and Gender Identity, 179–180
Staudt, Kathleen, 60–61
Stean, Jill, 99
stratification, 43
"structural functionalism," 38–40
subject formation: and gendered subjectivity, xxiii, xxv, 175; local processes of, 53–54; maternal, 30, 34, 37, 40, 47, 49, 51; men's processes, of 72–73, 75; modern, 183; queer ; women's processes of, 54, 72–73, 75. See also gender; sexuality
"substantive equality," 5
Suriname: matism in, 190, 218n15; women's reproductive health in, 113
Swarr, Amanda, 174, 178–179
"symbolic capture," 192–197, 202–203

T

"tailored mainstreaming," 56. *See also* Rees, Teresa
Tambiah, Yasmin, 218n12
termination of pregnancy: as human right, 96–97; illegality of in

Caribbean, 113–115; in Antigua and Barbuda, 113, 214n25; in the Bahamas, 113; in Barbados, xxv, 92, 113, 114, 116, 117, 119–122, 214n26, 215nn33–34; in Belize, 113; in Cuba, 113; in Dominica, 113; in Grenada, 113; in Guyana, xxv, 113, 114–116, 118, 119, 120–122, 214n22, 214n27; in Haiti, 113; in Jamaica, 113; in St. Kitts and Nevis, 113; in St. Lucia, xxiv, 92, 113, 124–125; in St. Vincent and the Grenadines, 113; in Suriname, 113; in Trinidad and Tobago, xxiv, 92, 113, 122–124, 125, 214n23; maternal mortality worldwide from, 100–101; maternal mortality from in Trinidad and Tobago, 111–112. *See also* reproductive rights

Thoreson, Ryan, 191

time: as category of analysis, 204 (*see also* historicization of gender); as resource, 144–145

timelessness, 183

Tindigarukayo, Jimmy, 157

transgender, 174, 190, 201–202, 217n3

transnational feminist critique, xxvi, 10, 11–12, 19, 91–92, 106–107, 108, 187–188, 190, 204. *See also* post-structuralism

transnational organizing, xxiv–xxv, xxvi, 10–11, 19, 98, 100, 102, 116, 186, 199, 200, 202–203, 207n8

"transnationalism reversed," 11. *See also* Friedman, Elizabeth

Trinidad and Tobago: common-law relationship recognition in, 74; domestic violence protection in, 73; economy of, 103; equal opportunity legislation in, 154, 216n8, 216n9, 216n10; GDI, GEM, and HDI ranking of, 70; gay rights (*see* sexuality/rights in); "gender dilemma" in, 79; gender empowerment in, 85; gender unit in (*see* Trinidad GAD); institutional mainstreaming in, xxiii, xxiv, 23–24, 89, 210n1, 210n7 (*see also* Trinidad GAD); labor riots in, 31–35, 208n11; maternal mortality in, 111–112, 122–123; national gender policy of, 89; nationalism and development in, 211n11; nationalism and gender equality in, 72–73, 110, 214n19; nationalism and sexual minority equality in, 184–185; sexual harassment and gender equality in, 154, 155, 156, 216n8, 216n9, 216n10; sexuality/rights in, 51, 188, 194, 218n12, 219n22; slavery in, 25–31; social services and welfare in, 24, 26, 46–54 (*see also* WIRC); within Anglophone Caribbean community, 205–206n1; women's identity as mothers in, 25–44; women's movement in, 73, 79, 85; women's reproductive health/rights in, xxiii, xxiv, 92, 95, 110, 111, 112, 113, 122, 123, 124, 214 n19, 214 n23 (*see also* ASPIRE)

Trinidad Colonial Office, 32

Trinidad Gender Affairs Division (GAD), xxiv, 55, 72–89, 122, 153, 211 n17 (*see also* IDB Non-Traditional Skills Training Program)

Trinidad Workingmen Association, 33

Trotz, Alissa, 135, 141–142

True, Jacqui, 6, 8–9, 125, 206n2

U

"underdevelopment," 11, 205n2

unisex, 201

United Gay and Lesbian Association of Barbados (UGLABB), 219n22

United Nations: and development (*see* HDR; MDGs); and gender rights advocacy, 177, 179–180; and sexuality rights advocacy, 178–180; Children's Fund UNICEF), 9–10; conferences for/on women, 8, 71, 98, 99, 205n3, 214n17 (*see also* Beijing; Cairo/ICPD; CEDAW; Nairobi FLS; Safe Motherhood Conference); Convention on the Rights of the Child, 180; Decade for Women 6, 73; Development Fund for Women (UNIFEM), 9, 63, 66, 210n1, 211n10, 215n1; Development Program (UNDP), 205n3, 210n6; Economic

248 Index

Commission for Latin America and the Caribbean (UNECLAC), 9, 113, 211n10; Economic and Social Council (ECOSOC), 7, 17, 179, 180, 217n9; Economic and Social Commission for Asia and the Pacific (UNESCAP), 214n17; Human Rights Committee (HRC), 98, 179, 180, 218n11; Universal Declaration of Human Rights, 177, 180, 189
United States Aid for International Development (USAID), 108
University of the West Indies, 62, 64, 202; Centre for Gender and Development Studies, 88–89; Distance Teaching Experiment, 64; Programme in Gender and Development, 88; Rights Advocacy Project (U-RAP), 202; Women and Development Studies Project, 62, 88–89

V
"variegated hostilities," 197, 201, 203, 219n27, 220n29
Veitch, Janet, 56, 210n4
"vernacularization" (of sexuality rights), 182, 183, 184, 186, 188–189, 218n11. See also speculative vernaculzarization; Yogyakarta Principles
visions (strategic development plans), xxv, 91, 211n11. See also MDGs
Vizenor, Gerald, 198

W
Wahab, Amar, 187–188
Waites, Matthew, 179, 181, 190
Wallace, Tina, 57–58
Walcott, Rinaldo, 188, 195
Wallerstein, Immanuel, 213n8
"wastage," 144–146, 216n5. See also RBPF
Weber, Max, 61, 153
welfare services, xxiii–xxiv, 12, 24, 25, 32, 35, 38–40, 44–54, 73, 87, 208n11, 209n17; in Trinidad and Tobago, 44–53
well-being as right/equity, 52, 96, 98, 102–104, 105, 106–108, 115, 125, 198, 211n15, 213n12
Wekker, Gloria, 190, 218n15
Wendoh, Senorina, 57–58

West India Royal Commission (WIRC), 32–36, 38, 208n13, 209n18
Westminster system of government, 8, 206–207n5
Whaley-Eager, Paige, 98–99
Wicks, David, 129
"woman" as category, 96
"woman on welfare," 53
woman's identity as mother, 107–108
Women and Development (WAD), 62, 79, 88, 126
Women in Development (WID), 21, 60, 67, 207–208n1
women's bureaus/desks/units in Caribbean, 62–63, 64, 65–66, 74; constraints on (see bureaucratic structure; illegality; incompetence); in Anguilla, 64; in Barbados, 64, 65; in Belize, 64, 65; in British Virgin Islands, 64; in Cuba, 64, 66; in Dominica, 64, 66; in Jamaica, 64, 66; in St. Kitts and Nevis, 64, 66; in St. Lucia, 64, 66; in St. Vincent and the Grenadines, 64. See also gender focal points; Trinidad GAD
Women's Movement, xxvi, 10, 12, 58, 65, 73, 104, 211 n16
women's rights as human rights, 89, 96, 99, 106, 177, 217n7, 218n11
Women's Second Chances Program (Trinidad), 84
Women Working for Social Progress (Trinidad), 73, 212n19
workers, women as, xxiii, xxv, 47
World Bank, 17, 210n6
World Health Organization (WHO), 100–101

Y
Yeboah, David, 108–109, 144
Yogyakarta Principles, xxv, 172, 180–192; limitations of, 181–184, 186, 189, 191–192; transformative possibilities of, 183, 186, 190–191, 199–200

Z
zami, 218n16
Zimbabwe African National Union-Patriotic Front Party (ZANUPF), 92
Zippel, Kathrin, 160